Planning Estuaries

NATO • Challenges of Modern Society

A series of volumes comprising multifaceted studies of contemporary problems facing our society, assembled in cooperation with NATO Committee on the Challenges of Modern Society.

Recent volumes in this series:

Planning Estuaries

Cees-Jan van Westen
Ministry of Transport, Public Works, and Water Management
Delft, The Netherlands
and
Reinier Jan Scheele
Utrecht University
Utrecht, The Netherlands

Published in cooperation with
NATO Committee on the Challenges of Modern Society

PLENUM PRESS • NEW YORK AND LONDON

Library of Congress Cataloging-in-Publication Data

van Westen, C. J.
Planning estuaries / Cees-Jan van Westen and Reinier Jan Scheele.
p. cm. -- (NATO challenges of modern society ; v. 20)
"Published in cooperation with NATO Committee on the Challenges of Modern Society."
Includes bibliographical references and index.
ISBN-13: 978-1-4613-8033-7 e-ISBN-13: 978-1-4613-0397-8
DOI: 10.1007/978-1-4613-0397-8
1. Estuarine oceanography. 2. Coastal zone management.
I. Scheele, Reinier Jan. II. Title. III. Series.
GC97.W47 1996
333.91'64--dc20 96-21923
CIP

This book represents the result of a pilot study on Estuarine management initiated by the NATO Committee on the Challenges of Modern Society

ISBN-13: 978-1-4613-8033-7

Softcover reprint of the hardcover 1st edition 1996

A Division of Plenum Publishing Corporation
233 Spring Street, New York, N. Y. 10013

10 9 8 7 6 5 4 3 2 1

PREFACE

The pilot study on estuarine management was initiated by the NATO Committee on the Challenges of Modern Society (CCMS). The objective of this project was to make a comparison of several estuarine management strategies and to formulate a set of recommendations and planning guidelines to restore polluted estuaries, to prevent further estuarine deterioration, and to manage conflicting demands stemming from human activities. The first results of this NATO-CCMS project were reported in 1982, describing six estuaries and their management approaches. In 1985 CCMS decided a subsequent research project should be performed to achieve the original objectives. This second research project was basically done by way of a comprehensive questionnaire that was sent to authorities of estuarine areas all over the world.

Eleven questionnaires have been received. In some instances, the questionnaires were not fully completed for several reasons. Especially, data on water quality was lacking. As a result, less attention will be paid to related aspects as originally intended.

This study shows that knowledge of the number of functions and conflicts of estuaries may provide relevant information concerning the severity of the problems as well as the propensity towards planning of water systems.

Therefore, as it is likely that investigations of this rather straightforward nature will not demand highly specialized skills, it is important to stimulate at least an investigation of functions and conflicts of estuaries in all suspect areas of the world. Such an investigation may provide the basis for a mega strategy concerning planning development. Priorities can then be set in order to improve the situation.

The report focuses particularly on policy development of estuaries. An extensive literature survey was carried out and the main results are presented in this report. Although in general a growing amount of literature is available on the subject of estuaries by now, little is produced on the issue of integrated policy planning for estuarine areas.

Connected to this is the lack of guidelines in making integrated policy plans. For this reason, thirteen guidelines have been proposed that may contribute to a more effective and comprehensive policy for estuarine areas:

1. Constituting an institutional framework.
2. Defining the boundaries of the area covered by the plan.
3. Assessing the existing situation and ways in which the area is currently being used.
4. Assessing the expected area development if no change of policy would occur.
5. Assessing the potentials of the ecosystem and the requirements of all the users.
6. Evaluating the use of the area, the changes in prevalent uses, and the ways in which these relate to other uses and factors; obtaining information on actual or

potential areas of conflict and identification of areas on which information is needed.

7. Formulating the aims and ways in which the uses of the area should be developed.
8. Making a selection from these options and deciding on the proper mode of management to achieve them.
9. Identifying the administrative and legal instruments required by the specified management mode, and developing them if necessary.
10. Charting the consequences of the general policy for the policies of the regulatory agencies active in the area.
11. Specifying in an action plan the objectives to be achieved by the agencies involved for particular stages of the planning period.
12. Formulating a procedure for modifying the plan.
13. And, all too often neglected: formulating a provisional interim policy to be implemented immediately as the estuary may undergo substantial changes during the planmaking period.

In particular, the near disaster situations experienced in 1995 on major river systems in Europe, again revealed the very necessity of a provisional as well as an integrated approach towards water management.

Cees-Jan van Westen
Reinier Jan Scheele

ACKNOWLEDGMENTS

Many persons contributed substantially to the making of this publication. Carel Colijn, Paul Stortelder and Maarten Knoester initiated the second stage of the NATO CCMS Pilot Study on Estuarine Management. Hans van Nunnikhoven conceived the questionnaire on estuarine management. Cees Groenewoud managed the copies returned. Water quality data was processed by Albert Holland.

Literature study was by Paul Francissen. The diagrams were produced by Jo de Brabander and Jan van den Broeke. Marijke van der Zande and Carla Schrader-van der Laan took care of typing the tables.

Notwithstanding these extensive contributions, only the authors are to blame for any incorrections that may have entered the final product.

The authors want to express their special gratitude to those who did fill in, or those who took responsibility for the filling in of the very extensive questionnaires:

Ebro, Spain	J.R. Vincent
Huelva, Spain	J.R. Vincent
Ave, Portugal	Mrs. A. Monzon Capape
Tejo, Portugal	Mrs. M. Cardoso da Silva
Loire, France	M. Chaussepied
Lagoon of Venice, Italy	A. Puddu
Solent, United Kingdom	A.P.M. Lockwood
Eastern Scheldt, The Netherlands	R.C. Boeije
Western Scheldt, The Netherlands	R.C. Boeije
Cumberland, Canada	D.C. Gordon
Tcacapan, Mcxico	V.A. Fuentes
Information of Denmark	B. Sogaard

CONTENTS

1

INTRODUCTION

1.1. PILOT STUDY ESTUARINE MANAGEMENT

In the twentieth century industrialization significantly added to the advantages of coastal areas for the location of human activities. This resulted in a concentration of human activities and urbanization of coastal areas. In the last decades development in these areas has intensified as a result of the boom of the tourist industry. These growing demands on coastal resources place serious threats for coastal areas as evidenced by a progressive deterioration of the coastal environment (Coccossis, 1985, p.21).

In this process estuarine areas seem to be threatened extra. Estuaries are convenient places in which to dump wastes, they may be naturally good sites for ports and docks, while their shores present possibilities to reclaim land for commercial and residential developments. In addition to such activities on the shores of estuaries themselves, their location at river mouths rends them susceptible to the effects both of wastes dumped upstream and of interference with the flow of fresh water into them. Estuaries are also more susceptible to the results of human activities than are the waters of open coasts. This applies particularly to the discharge of harmful materials since in estuaries these are not dispersed nearly as rapid as in the open ocean (Allan, 1983, pp.2-3).

According to McLusky (1989, p.177) the management of estuaries has to cope with three fundamental paradoxes: firstly, the majority of the world's major cities are located alongside estuaries, yet for most of the inhabitants of those cities, estuaries are the most natural wildlife habitat that they encounter. Secondly, most of the major estuaries of the world are to some degree polluted; yet, in many countries more estuarine areas are designated nature reserves than any other habitats. Thirdly, many estuaries are organically enriched, yet all estuaries are amongst the most productive natural ecosystems known.

Estuaries are important to man. Not only economic features but also environmental and social issues force a growing interest into the well-being of estuaries. The costs of deterioration of water systems express themselves more and more in hard figures. For example the algae blooms in the Adriatic Sea in the summer of 1989 resulted considerable financial loss due to the massive decrease in the number of tourists. The loss of nursery grounds of fish in estuaries demonstrates not only the decline of environmental values, but involves tangible economic effects on fisheries, too. Therefore, sound and balanced management of estuaries is more and more considered as a basic need of modern society.

For this reason a research project was initiated by the NATO Committee on the Challenges of Modern Society (CCMS) as long ago as October 1978 to look into this problem. The original proposal for a pilot study was prepared by a group of experts meeting in Athens in January 1979, and was accepted at the plenary sessions of CCMS in 1979. The original objectives of the study were stated as follows:

Figure 1. Coastal waters attract people.

Figure 2. Industrial activities often concentrate along the borders of estuaries. The dumping of wastes leads to a serious deterioration of the environmental quality.

1. Compilation of case histories of various approaches to estuarine management in the participating countries.
2. Analysis of the effectiveness of these management strategies.
3. Development of a set of recommendations and planning guides to assist member countries in choosing the most effective management tools and strategies for their particular conditions.
4. If possible, the development of a plan for the environmental management of a selected estuarine area, which could be implemented as a CCMS demonstration activity.

In May 1982 a final report (CCMS, 1981) was submitted to CCMS. Considering this report it can be concluded that only the first objective was achieved. Six estuaries and their management approaches were described at varying levels of detail. The effectiveness of management strategies, objective number two, was superficially dealt with, due to time constraints and availability of information. Objectives three and four, a general planning guide and a plan for the environmental management of a selected estuarine area, did not materialise at all.

In 1985 CCMS decided a subsequent research project should be performed to achieve the original objectives. These objectives were slightly rephrased and may be stated as follows:

1. the review of existing problems and management strategies based on existing literature, and information provided by the member countries.
2. the analysis of differences and resemblances in both estuarine problems and management strategies and the effectiveness of these strategies;
3. the development of a set of recommendations and planning guidelines to assist member countries and other participants in choosing the most effective management tools and strategies for their particular conditions regarding present problems as well as preventing potential problems.

This report presents the results of an analysis of estuarine problems and management strategies and contains a set of recommendations and guidelines.

1.2. PLAN OF ACTION

A comprehensive questionnaire has been sent to appropriate authorities of 25 countries all over the world. The topics involved concern:

- general geographical and ecological features
- functions and conflicts in the estuaries involved
- policies intended in relation to estuaries
- research and management undertaken

Eleven questionnaires have been returned, including substantial information on the topics mentioned. The filled-in questionnaires cover the estuaries mentioned in Table 1.

Besides these estuarine areas, related information was sent by Portugal, Greece and Denmark concerning, among others, planning, biological data and maps.

Some problems arose in processing the questionnaire data. The amount of data in some of the questionnaires was rather restricted; especially data on pollution was lacking. Some of those who completed the questionnaire were specialised in one of the areas concerned, but lacked information on other issues; evidently, a biologist might have problems with planning issues and a planner vice versa. Subsequent contacts with additional

Table 1. Estuaries covered by questionnaires

Estuary	Country
Ebro	Spain
Huelva	Spain
Ave	Portugal
Tejo	Portugal
Loire	France
Lagoon of Venice	Italy
Solent	United Kingdom
Eastern Scheldt	Netherlands
Western Scheldt	Netherlands
Cumberland	Canada
Teacapan	Mexico

experts resolved most of these problems. Still the questionnaires and related reports and books didn't provide enough information to formulate a complete set of guidelines for estuarine management. For this reason a literature review was carried out and also a meeting of Dutch experts was held in Middelburg in October 1993. During this meeting recommendations were drafted and a set of guidelines was completed.

Considerations regarding the actuality and partial availability of water quality data brought about a shift towards an emphasis on literature study and the development of guidelines.

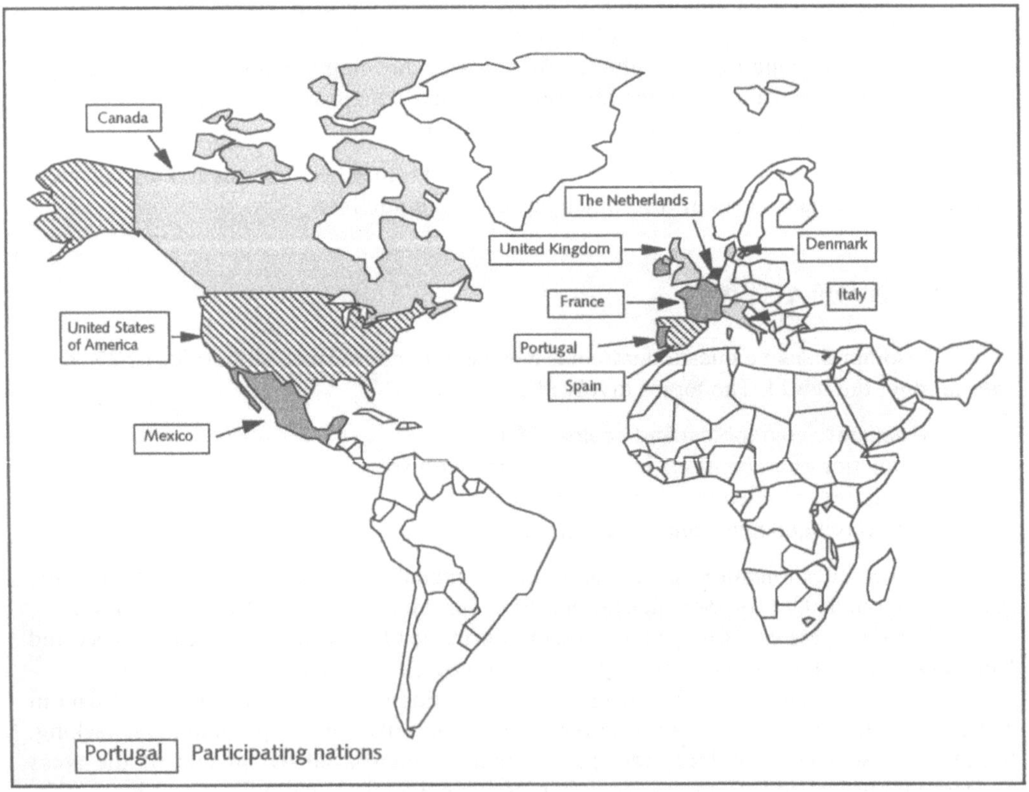

Map 1.

1.3. DEFINITIONS

Definition of an Estuary

An estuary is defined as follows:

> An estuary is a semi-enclosed coastal waterbasin with an open, possibly controllable, connection with the sea and supplied with fresh water from the land. This body of water is more or less affected by tidal action and within it seawater is measurably mixed with and diluted by fresh water from the land and rivers.

An estuary is an ecosystem in which many specific and strongly related physical, chemical and biological processes occur. Its landward boundary lies where the influence of the tide becomes noticeable. Towards the sea the boundary lies in a zone where a transition occurs between the area with rapid and relatively large fluctuations in temperature of the interstitial water, and the area where the tidal geomorphology of channels, intertidal flats, and shallow waters, changes into the normal ocean-floor (Saeijs, 1982, p. 36).

Definition of Estuarine Management

In this study on estuarine management attention will be paid to the various management aspects which have to be considered while improving the situation of estuarine areas. For that purpose it is first of all important to define more clearly what estuarine management is. In most literature on estuarine development the expression "estuarine management" is used to indicate the human interference with physical, chemical or biological processes in estuaries. Attention goes out to the technical aspects of measures to control these processes, to the environment in which these processes take place and to the consequences of the measures being taken. Estuarine management is thus considered as the whole of activities which are of direct significance to the state of the estuary.

In this report the expression estuarine management is used in an other sense. Here, estuarine management does not indicate the activities of direct interference, but indicates the way in which activities are organized. Management is conceived as the process of getting activities completed efficiently with and through other people (Robbins, 1988). To this definition four major management functions can be distinguished:

- Planning; defining management objectives and developing a strategy to reach these objectives;
- Organizing; designing an organizational framework;
- Leading; coordinating the activities;
- Controlling; monitoring the activities and if necessary carrying out adjustments.

Estuarine management thus is part of "coastal resource management" as defined by Hildebrand (1989, pp.9-10). He distinguishes two components to this definition: planning and management. Integrated planning is a process designed to interrelate and jointly guide the activities of two or more sectors in planning and development. The goal of integrated planning is the preparation of a comprehensive plan which specifies the means to effectively balance environmental protection, public use and economic development to achieve the optimum benefit for all concerned. The integration of activities usually involves coordination between data gathering and analysis, planning and implementation.

Coastal management is the process of implementing a plan designed to resolve conflicts among a variety of coastal users, to determine the most appropriate use of coastal resources, and to allocate uses and resources among legitimate stakeholders. Management is the actual control exerted over people, activities and resources.

Recent Development in Estuarine Management

In the last decade a changing attitude towards estuarine management emerged. In general, water management used to be directed towards the protection of land against the water and towards keeping sufficient water available for the various user groups. In most cases water quality management only tried to cope with troublesome visible and smelly pollution of surface waters. Attention for the natural environment as such is more recent, as is the objective to create a healthy aquatic environment. An example of this is found in the Western Scheldt estuary (Netherlands), where management used to be restricted to, for instance, dredging shipping lanes and harbours, maintaining dykes and sluices, beaconing the navigable passages etc. Nowadays, the estuary is managed as a multi-functional system in which issues such as water quality performance and natural values are also taken into account (Scheele, 1991, pp.199-201). As a result of the growing awareness of the worsening of the environmental quality, the ecological aspect has become a major item in water management during the last ten years. Attention can no longer be restricted to water problems only. Water has to be seen in relation to its physical environment: surface water and its bottom, its banks, and its related surrounding. Besides, water management can not be considered as an isolated activity. The great number of relations with other policy fields is taken more and more into consideration by managers involved (Glasbergen, 1991, p.240).

According to Smith (1991, p.273) we are witnessing the beginnings of an approach to the development and management of the world oceans and seas based less on individual uses, state and industrial interests, and more on relationships among these, as these relationships become progressively more important.

This reoriented approach to water management and estuarine management clearly fits in the concept of sustainable development, which is described in the Brundtland Report Our Common Future (1987) as "development that meets the needs of the present without

Figure 3. Waves, wind and currents constantly change the patterns of channels.

compromising the ability of future generations to meet their own needs." The set of recommendations and guidelines in this report are formulated to give a concrete form to the concept of sustainable development for estuarine management.

1.4. GUIDE FOR THE READER

In Chapter 2 the geographical characteristics of the estuaries involved will be considered. Current and future problems originating from conflicting functions in the area, are taken into account. Furthermore, a classification is conceived of water pollution, regarding general parameters, inorganic and organic micropollution and oil. This study aims at presenting a rank order of pollution conditions of the estuaries concerned. This rank order should provide for a better understanding of a statement by Saeijs (1982, p. 392), hypothesizing that management strategies differ in severely polluted water systems and relatively clean ones.

Also the management strategies for these estuaries will be compared and discussed. They will be classified according to the severity of pollution and the number of functions and conflicts that have been noted.

The effectiveness of various policies is hard to judge, especially in those cases where no clear objectives are incorporated in the plans. However, the objective of diminishing pollution is evidently present in all of the longer ranging management projects. Other issues, such as the cooperation of public authorities, laws and measures involved, as well as the availability of research programmes are also taken into account. These issues have proved to be of vital importance in earlier projects.

Chapter 3 is a literature review on estuarine management in order to get insight in the present knowledge on this subject. Although it was expected from the very start that the number of publications would be limited, the study was done in order to prevent existing material from being ignored in the course of the research project.

Subsequently, in Chapter 4 recommendations and planning guidelines are formulated. Different pollution levels and numbers of conflict pose different conditions to management. Furthermore, political and democratic structures of a country may influence choices for central or decentral options of control. However, some planning issues have shown to be indispensable in general. The guidelines and recommendations presented in this report should not be interpreted as strict rules; the aim is to formulate minimal contents of a plan, proposals for decisions to be taken and recommendations regarding issues that need special attention. A list of literature is presented in Chapter 5.

2

CHARACTERISTICS OF ESTUARIES

2.1. INTRODUCTION

This chapter describes the estuaries involved in the pilot study and the strategies developed for the management of these areas. Firstly, the geographical characteristics will be presented. Whenever possible, the information is presented in a map or table, enabling the reader to obtain an overall picture. Also the scale of pollution, the presence and intensity of functions and the relevant conflicts now and in the future will be highlighted.

For several estuarine areas management plans have been drafted to set policy objectives and to present management measures to achieve these. The plans will be described and subsequently some conclusions will be drawn.

2.2. GEOGRAPHICAL CHARACTERISTICS

The estuaries described demonstrate many differences. The jawning funnel of the Western Scheldt opposite the small Ave river mouth; the city of Venice in contrast to villages such as Amherst and Sackville situated on the Cumberland Basin in Canada. Tables 2 and 3 illustrate some geographical aspects. For instance, mangrove swamps in the lagoon of Teacapan and the salt marshes and pebble beaches of the Solent; the mean tidal differences of 11 meters of the Cumberland and the 1.1 meters of the Teacapan. Tables and maps in this part are self-evident. Hence, we will confine ourselves to a short description of the areas.

Ebro

The Ebro flows from the north-west of the Iberic Peninsula to the south-east. The river discharges in a delta at the Mediterranean Sea. The delta area is situated in the southern part of Catalonia and is mainly used for agricultural purposes (rice, vegetables). This has induced a negative effect on the habitat of flora and fauna. The lagoons and marshes are known for their abundant birdlife. About 65,000 birds reside permanently in the region. Of old times, bownet-fishing has been practised in the lagoon, and the inhabitants still catch some 250,000 kg of fish per year. The draining of sewage water, loaded with industrial nutrients in particular, causes problems for aquaculture and tourism. Also, the pollution of upstream cities contributes to the eutrophication. As this situation continues, problems will get worse in the future.

Huelva

The estuary of the rivers Odiel and Tinto is situated in the south of the province Huelva in the south-west of Spain, near the Portuguese border. It debouches into the Atlantic Ocean.

Table 2. Geographical and physical aspects

	Estuary										
	Ebro	Huelva	Ave	Tejo	Loire	Venice	Solent	Eastern Scheldt	Western Scheldt	Cumberland Basin	Teacapan
Geographical Aspects											
Length (km)	26	21	2	50	95	52	54	40	75	51	131
Width (km)											
Min	16	3.5	0.069	0.4	0.225	8	0.5	3	2.2	1	0.1
Max	38	14	0.2	12.6	3	14	5.5	10	5	4	7
Shorelines (km)											
Rock/cliff				11			6.5			15	
Dunes	90					3					
Dikes		13			80	13.5		125	210	50	
Beach			9	9.5		22.5				35	95 (sand barrier)
Other				116.5 (marsh)	30 (wetlands)		33 (mud) 62 (gravel) 6.5 (shingle)				
Population Estuarine Coast ($*10^3$)	50	153	21	2000	557	708	270	30	80	20	258
Physical Aspects											
Long-term mean tidal difference (m)											
Fresh water limit				3.3	4		3.03		5.0	11	
Base line sea				2.8	4	1.40	2.83	3.30	4.0		0.9
Max. tidal velocity (m/s)											
Head				1.2	0.7	0.3	0.98		1.2		
Midpoint				2.0	1.8		1.9	1	1.2	1.5	0.3
Seaward				2.3	1.25	1.0	1.2		1.2		
Mean tidal volume (m^3*10^6/tide)		21.3		750	200	230	405	850	1100	1025	
Freshwater input (m^3/s)											
Mean	1.9	15.2	no	350	825	31	17.8	25	120	83	159
Winter	0.56				1300		26.2		180	54	

Waterdepth class (%)											
0 - 2.5 m		80		86	30	75	26	40	22		98
2.5 - 5 m				11	30	20	12	40	13		
5 - 20 m		20		3	40	5	52	10	52		2
> 20 m		.					10	10	13		
Water area (km^2)											
Mean high water (MHW)	19.4			340	66.2	523	130.2	350	290	124	1406
Mean low water (MLW)				300	35.3	432	121.5	225	200	70	273
Soil characteristics	muddy		mud	mud	sand	mud	mud	sand	sand	gravel	mud
			gravel	rock	mud	sand	sand	mud	mud	bedrock	
				sand			gravel				
Geomorphology (km^2)											
Intertidal area					31	91	102.8	104	94	54	1133
Area above MHW					3.5	44		6	34		4621

Table 3. Climate

Month	Jan.	Feb.	March	April	May	June	July	August	Sept.	Oct.	Nov.	Dec.	Prevailing wind	Force (m/s)
Ebro	9.5	10.7	12.3	15.4	6.9	22.1	24.9	25.1	22.5	18.0	13.3	10.1	N-NW	
	? 29.1	29.1	36.0	42.9	67.6	46.7	19.6	35.8	83.5	83.5	44.8	56.1		
	157	151	237	250	207	239	318	307	209	159	115	173		
Huelva	11.8	12.5	14.5	16.8	19.3	22.5	25.4	25.4	23.4	19.6	15.3	12.4	S-W	
	8.4	6.5	6.8	4.1	2.7	1.4	0.5	2.4	1.5	5.6	6.5	8.2		
	183	117	249	194	282	330	343	356	239	270	144	157		
Ave	7.5	10	12	14	15	17	20	20	17	15	12	10	N-E	
	15	10	15	7.5	7.5	5	2.5	1.0	5	10	15	15		
				220			300		200			120		
Tejo	10.8	11.6	13.6	15.6	17.2	20.1	22.2	22.5	21.2	18.2	14.4	11.5	N	4
	11.1	7.6	10.9	5.4	4.4	1.6	0.3	0.4	3.3	6.2	9.3	10.3		
	–	–	–	–	–	–	–	–	–	–	–	–		
Loire	5.6	6.7	7.9	9.8	13.5	16.8	19.0	18.8	16.7	12.3	8.6	6.0	S-W/W	5-6
	8.2	7.7	6.5	4.7	5.6	3.3	5.4	4.9	5.4	5.6	9.3	7.0		
	69	91	154	182	198	245	264	241	207	133	96	70		
Venice	2.8	4.8	7.8	12.1	16.7	20.4	22.4	22.0	18.7	13.6	8.3	3.1	N-E	5
	6.0	5.8	5.5	7.2	7.3	7.0	6.2	8.7	5.9	7.2	10.0	4.8		
	–	–	–	–	–	–	–	–	–	–	–	–		
Solent	4.7	5.0	6.8	9.3	12.5	15.5	17.2	17.1	15.0	11.9	7.8	5.9	S-W	3
	8.2	5.9	4.5	4.5	5.5	5.5	4.9	6.8	7.7	7.4	8.8	8.4		
	56	71	122	165	211	214	207	191	152	112	73	53		
Eastern	3.1	3.1	5.2	8.0	11.9	14.9	16.7	17.0	15.2	11.7	7.2	4.5	S-W	5-6
Scheldt	6.1	4.9	4.6	4.4	4.4	5.8	7.1	7.7	6.7	7.3	7.7	7.0		
	50	67	117	165	210	217	201	191	151	107	56	42		
Western	3.1	3.1	5.2	8.0	11.9	14.9	16.7	17.0	15.2	11.7	7.2	4.5	S-W	6
Scheldt	6.0	5.1	4.6	4.4	4.5	5.8	7.1	7.7	6.7	7.3	7.7	7.0		
	50	70	118	165	210	217	201	191	151	107	56	42		
Cumberland	-1.5	-1.5	1	4	9	13	17	20	15	10	5	1	W	5-10
	–	–	–	–	–	–	–	–	–	–	–	–		
	–	–	–	–	–	–	–	–	–	–	–	–		
Teacapan	21	21	22	25	27	30	30	29	28	28	26	24	W	1
	2	1	0	0	0	2	27	21	34	6	1	0		
	240	240	300	360	360	360	360	360	320	300	240	240		

22.1 mean temperature; 46.7 precipation in cm; 239 sun hours.

The original water system has silted up largely, so that marshy islands and channels were formed. The inner delta is a formation in evolution since the beds that have not been maintained were lost by silting up. The mines of Tanis and Rio Tinto have a large impact on the area, by the construction of metal piers for the transport of iron ore on one hand and by the pollution of the water on the other hand. The rinsing of the ore gave the beds a distinctive colour, from which the name Tinto, meaning coloured, originates. The city Huelva has seen a large growth from the fifties onwards. The petrochemical industry in particular has settled here and has given an impulse to the port activities. An outer harbour was developed, where deep draught ships can moor. In 1974 a long outside dike was constructed which reduced the filling up of the estuary with sand by ocean currents.

In this area, the conflicts between industry, mining, tourism, agriculture and dredging, are clearly defined. Plans for the reorganisation of industrial draining were drawn up, but have not been effective.

Ave

The river Ave is situated in the north-western part of Portugal, mostly in the Brags district. The estuary, which has a length of 94 km, runs through the Porto district from north-east to west. It rises in the Cabreira mountains at a height of 950 metres and has a drainage-area of 1,390 km^2.

Agriculture and industry are considerably well represented around the rivermouth and upstream. Especially the textile and fish industries are most important.

Vila do Condo is the largest centre in this area populating 20,630. There is an ornithological reserve with fir vegetation in the dunes near the mouth. The development of tourism threatens this nature reserve by disturbing the birds and by the competition for space. These problems also occur in the areas where tourism, nature and sewage discharges of the villages and industries meet. It is expected that conflicts will increase in the future, because the number and intensity of different uses of the estuary will increase.

Tejo

The Tejo is the largest river of the Iberic Peninsula, with a drainage area of 80,146 km^2 and a length of 875 km. It rises in the Serrania de Cuenca, lying to the east of Madrid. Near Lisbon the river flows into the Atlantic Ocean. The estuary entails a large bay, the "Mar da Palha," with an average depth of 7 metres and a narrow passage to the ocean. This corridor has a maximum depth of 46 m and has strong tidal movements. The estuary extends into the sea in the form of a submarine valley to a depth of 2000 m, which is a relict of earlier times. The northern part of the bay, with its vast mudflats, is a nature reserve. These intertidal areas are particularly known for their avifauna. These areas are also important for the function of fishing nursery. Industry, shipping, fishery and tourism have a large influence on the area. A wide-spread consensus exists on the necessity of the preparation of a management plan for the entire estuary.

Loire

The Loire has a water basin of 120,000 km^2 and debouches into the Atlantic Ocean. The agglomeration of Nantes and St. Nazaire, with half a million inhabitants, is the largest urban concentration on the border of this estuary. Shipping, harbour and industrial areas are the main components in the region. The harbour and industrial areas cover a territory of 3000 hectares. Tourism stimulates the economy of the area. The coastal hotels and pensions can house 200,000 persons.

The fishery trade employing some 300 fishermen, has e.g. for one species (elver) an annual turnover of 30 million francs. The agricultural activities are mainly confined to pasture and are of minor importance from an economic point of view.

The biological diversity is striking in the middle part of the estuary. The wide variety of plankton species and the strong benthos production are the obvious indicators of this variety. The nursery function for fish such as flounder, sole, whiting, seaperch and shrimp is concentrated in this water as well. The wetlands spread over an area of 40,000 hectares, notably to the north of the Loire, and offer a habitat for a large number of birds.

Problems occur with the pollution of the soil of natural areas abut the river. Water pollution has rather severe consequences for several functions. The ecologically valuable riverbanks are menaced by potential mud depots, too. Already, more than 100 ha of this territory is developed as mud depot.

Lagoon of Venice

The Lagoon of Venice is situated in the north-eastern part of Italy. The drainage area of the rivers Zero, Marzenejo, Naviglio, Brento and Deze, the tributaries of the lagoon, cover 2200 km^2 of land. The lagoon is a shallow saltwater system of 50 by 10 km. The connection with the Adriatic Sea consists of three wide openings. The current surface is only 70% of the surface in the beginning of the nineteenth century. Land reclamation for the benefit of agriculture, aquaculture, industrial areas, a road and an airport has caused this decline.

The water is salted up by the deepening and widening of the three outlets and the decrease of freshwater influx.

Occasionally, an algal bloom can be observed, brought about by the water pollution and the natural eutrophic climate. Contaminated waste water produces this effect and has disastrous consequences for fishery, aquaculture and tourism on Lido beach.

Solent

The rivers Test, Itchen and Hamble flow into the Southampton Water which in turn opens into the Solent. The latter, which also receives inputs from the smaller rivers the Beaulieu, Medina, Myon and Lymington, discharges on either side of the Island of Wight. Overall the Solent has a drainage-area of 3000 km^2 and a low freshwater input. The region is dominated by the sea ports of Southampton and Portsmouth. On the westbank of the Southampton Water lies an industrial area which accommodates one of the larger refineries in Europe and a power station. There is a fishery for oysters in the Solent and clams were taken commercially from Southampton Water. However, the major decline in this clam fishery resulted in the near cessation of clam dredging clearly improving the macro-benthos. Water sports are dominated by yachting and there is also public access to many beaches. Water pollution is not considered to constitute a significant problem except in localised regions associated with the output of water purification works and industrial plants. Southern Water has replaced the sewage outfall at Pennington and improved the treatment. Other rather minor changes concern the local rebuilt of sea walls and extension of fishery policing.

Next to a number of nature reserves, several areas around the Solent are designated as being sites of Special Scientific Interest indicating their ecological importance. A proposal of building a container port at Hythe could impinge on such a site.

Eastern Scheldt

The Eastern Scheldt forms a part of the Rhine-Meuse-Scheldt Delta. In former times the water of the Scheldt was carried off into the sea by way of the Eastern Scheldt. Gradually the Western Scheldt took over this function.

As a consequence of the Delta Plan no rivers flow into the Eastern Scheldt nowadays. This plan was designed after the flood of 1953. Primarily, the Delta Plan aimed to safeguard the south western part of Holland. This implied that all outlets to the sea had to be closed by dams except for two (New Waterway and Western Scheldt) because of navigation purposes. With the damming up of the Eastern Scheldt a large freshwater lake would have developed and the natural characteristics would have changed completely. Also the salt water fishery would have disappeared. Due to campaigns of environmentalists and fishermen a change in policy led to the construction of a storm surge barrier which only closes at very high waterlevels. In this way, the influence of the salt water and the tides would be preserved for the most part. In 1986 the storm-surge barrier became operational. The tidal range diminished from 3.70m to 3.30m, the total area from 45,000 ha to 35,000 ha. The morphology is characterised by great diversity. Intertidal areas and salt marshes, deep and shallow waters mark this estuary. The flora and fauna in, on and along the water show the same diversity. This estuary too has a nursery function. The water in the river is hardly polluted. Problems mostly occur at the interaction of nature, fishery and recreation.

Western Scheldt

The Western Scheldt is the last open estuary in the delta of the south western part of the Netherlands. The river Scheldt rises in the north of France, passes through the cities of Ghent and Antwerp in Belgium, and debouches into the North Sea. The greater part of the estuary is situated in Holland. The water basin covers 22,000 km^2 of which 14,000 km^2 in Belgium.

Navigation is the main function of the Western Scheldt; the estuary is the gateway to the ports of Antwerp, Ghent, Terneuzen and Vlissingen (Flushing). Hardly purified waste water is discharged into the river by industry and adjacent cities upstream. Therefore, at the border of Belgium and the Netherlands, the water has a low oxygen content and an excess of nutrients. This pollution lays a heavy burden on the natural functions of the intertidal flats and salt marshes.

Dredging of the Western Scheldt has a disturbing effect on the ecological system too. Recently, a decision has been made with regard to a further deepening of the shipping channel to allow 48' draught of vessels to the Port of Antwerp. Projects are developed to counterbalance the effects on the environment, especially the erosion of salt marshes and mud flats.

Cumberland Basin

The Cumberland Basin is a macrotidal estuary located at the head of Canada's Bay of Fundy between the provinces of Nova Scotia and New Brunswick. Several small rivers stream into it including the Herbert, Maccan, Nappan, La Planche, Missaquash and Tantramar. The mean tidal range is 11 m. About half of the total area is composed of saltmarshes and intertidal mudflats. The surrounding region is sparsely populated by 20,000 people, most of whom are concentrated in two towns near the head. Agriculture is the chief means of subsistence. Some forestry and mining also takes place. Much of the original salt marsh was diked by Acadians during the period of European settlement and most of the dikes remain today. The environmental impacts of this extensive diking have never been determined. While shipping and fishing were important local industries in the nineteenth century, the Basin is not used much today. The flora and fauna in and on the salt marshes, mudflats and turbid waters are subjected to little human disturbance. However, there have been serious proposals to construct a tidal power station for the generating of electricity at the mouth of the Basin which would cause major environmental changes. At the present time, such a project is not economically viable and there are no plans for construction, but this status could change in the future.

Map 2.

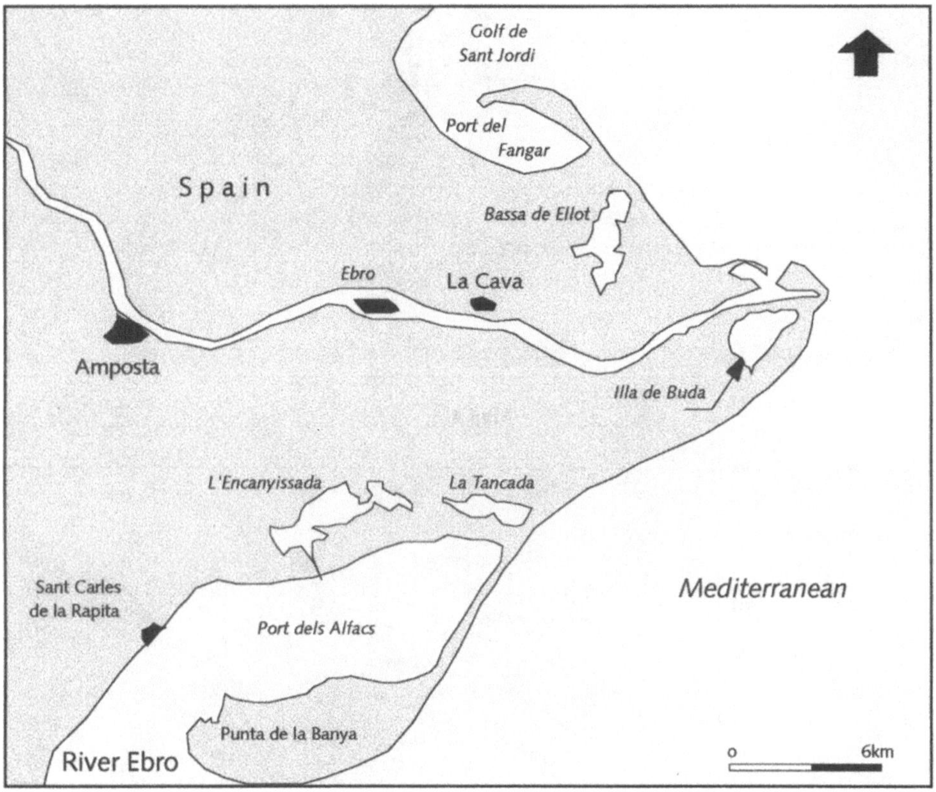
Golf de
Sant Jordi
Port del
Fangar
Spain
Bassa de Ellot
Ebro
La Cava
Amposta
Illa de Buda
L'Encanyissada
La Tancada
Sant Carles
de la Rapita
Mediterranean
Port dels Alfacs
Punta de la Banya
River Ebro
0
6km

Map 3.

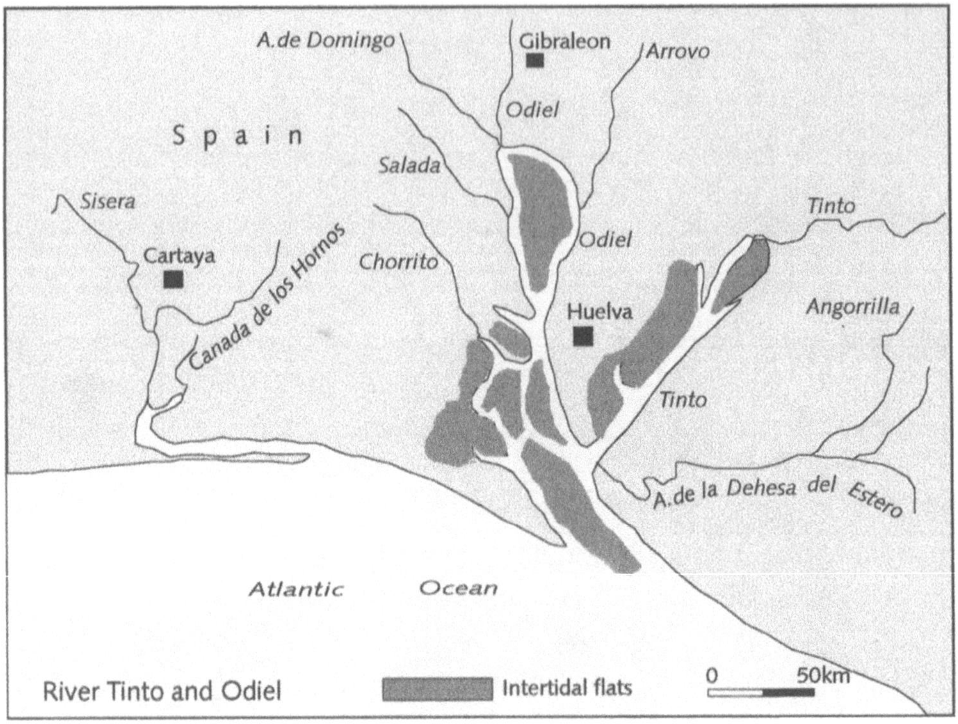
A.de Domingo
Gibraleon
Arrovo
Odiel
Spain
Salada
Sisera
Tinto
Odiel
Cartaya
Canada de los Hornos
Chorrito
Huelva
Angorrilla
Tinto
A.de la Dehesa del Estero
Atlantic Ocean
River Tinto and Odiel
Intertidal flats
0 50km

Map 4.

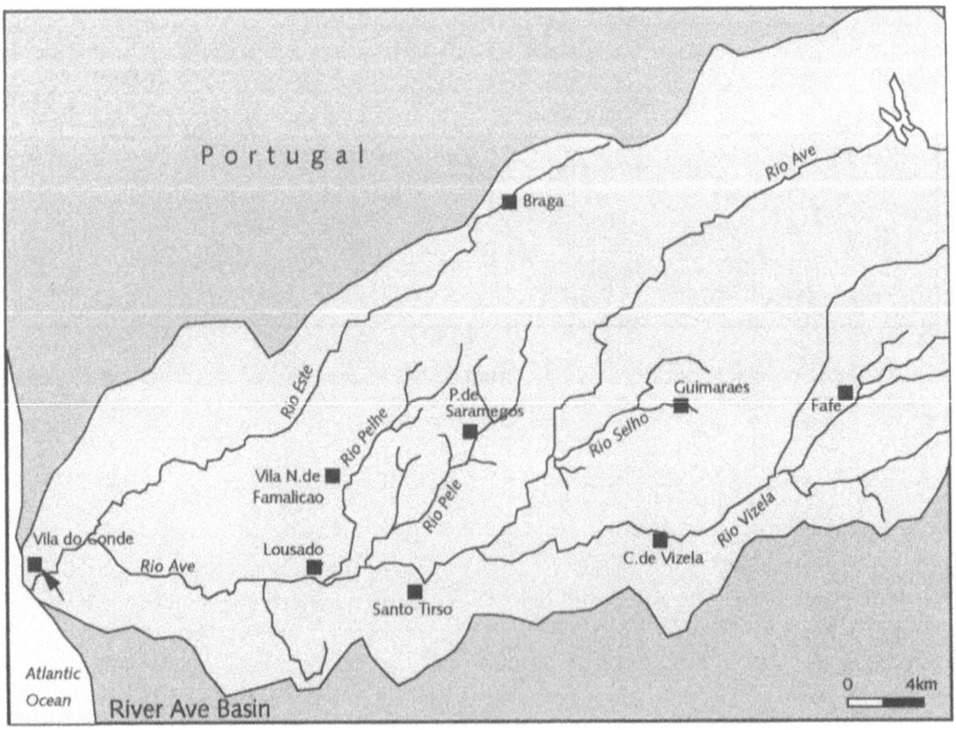
Portugal
Rio Ave
Braga
Rio Este
P.de Saramegos
Guimaraes
Fafe
Rio Pelhe
Rio Selho
Vila N.de Famalicao
Rio Pele
Rio Vizela
Vila do Conde
Lousado
Rio Ave
C.de Vizela
Santo Tirso
Atlantic Ocean
River Ave Basin
0 4km

Map 5.

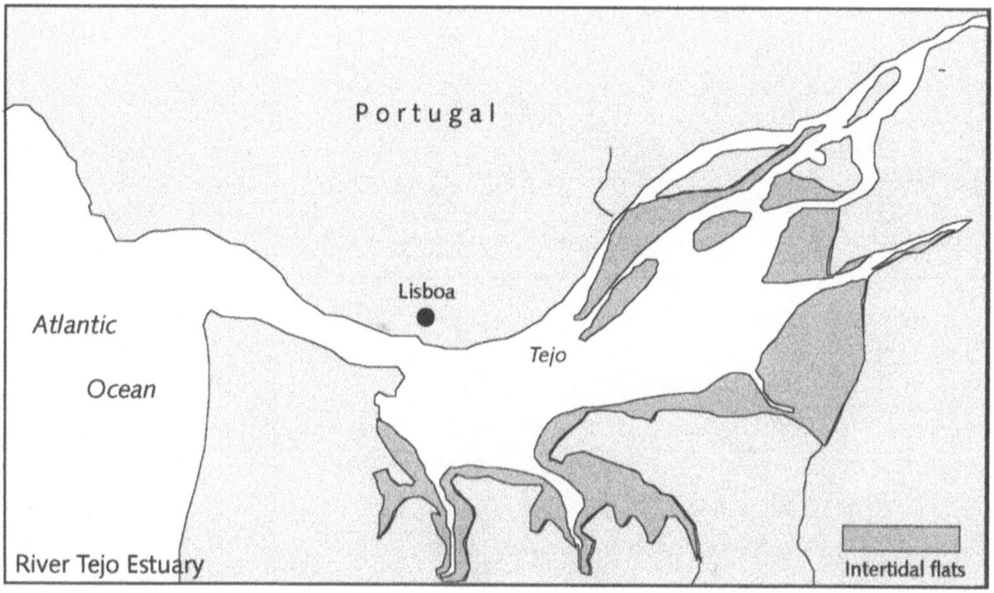
Portugal
Lisboa
Atlantic
Ocean
Tejo
River Tejo Estuary
Intertidal flats

Map 6.

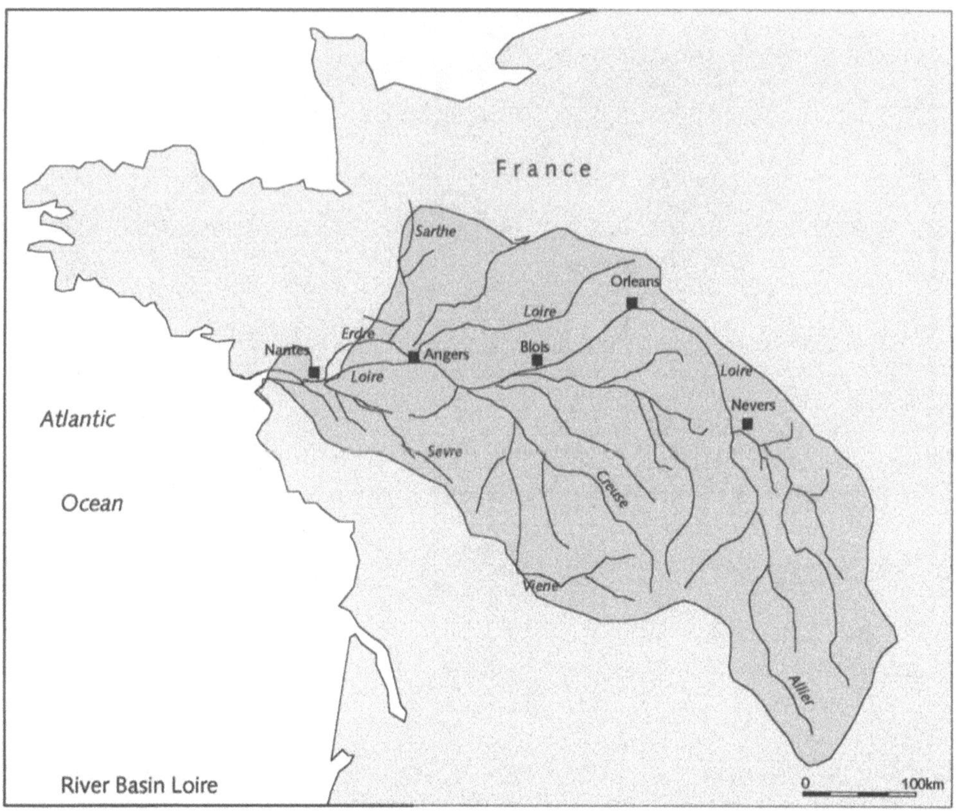
France
Sarthe
Orleans
Loire
Erdre
Nantes
Angers
Blois
Loire
Loire
Nevers
Atlantic
Ocean
Sevre
Creuse
Viene
Allier
River Basin Loire
0
100km

Map 7.

Map 8.

Map 9.

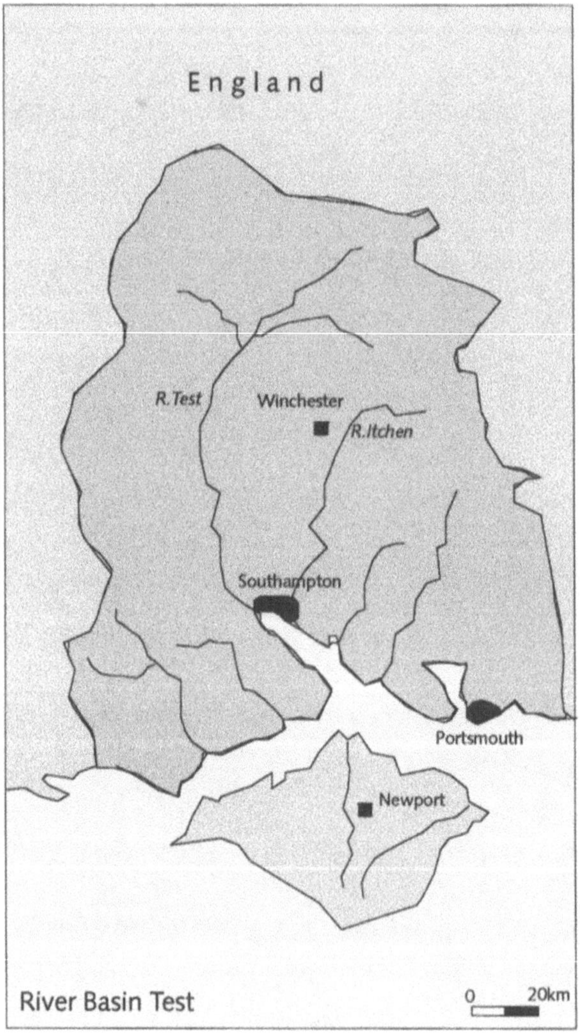
England
R.Test
Winchester
R.Itchen
Southampton
Portsmouth
Newport
River Basin Test
0
20km

Map 10.

North Sea
Amsterdam
The Netherlands
Eastern Scheldt
Western Scheldt
Germany
Belgium

Map 11.

Canada
Cumberland
Basin
United States of America
Atlantic
Ocean

Map 12.

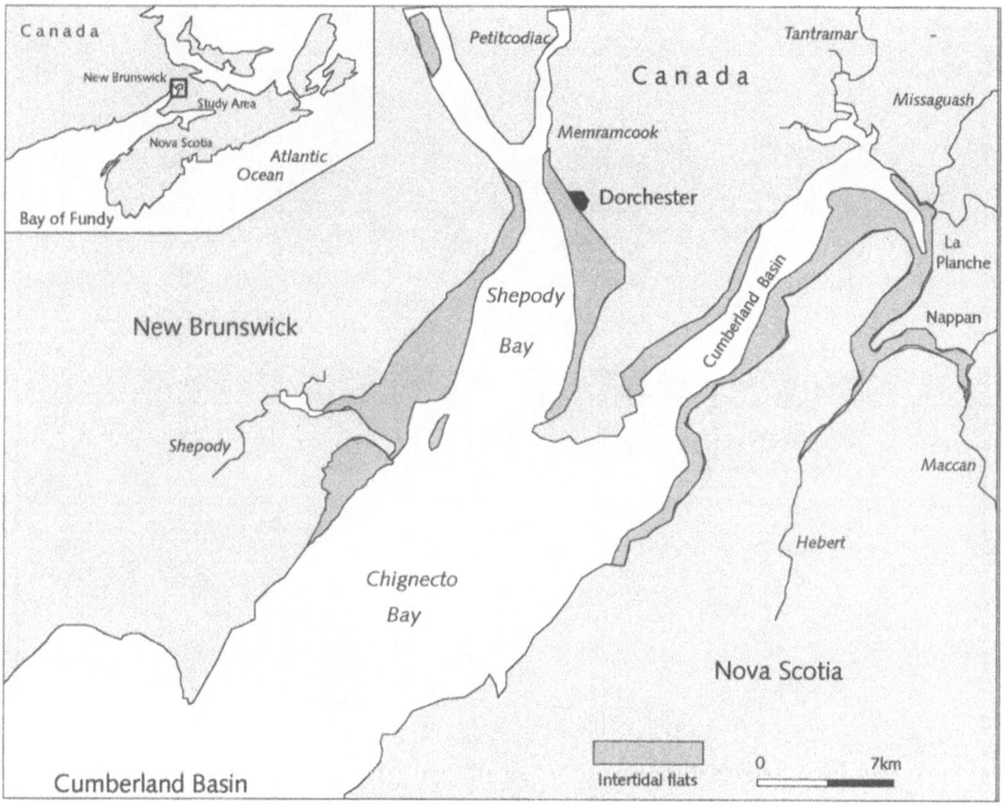
Canada
New Brunswick
Study Area
Nova Scotia
Atlantic
Ocean
Bay of Fundy
Petitcodiac
Tantramar
Canada
Missaguash
Memramcook
Dorchester
La
Planche
Cumberland Basin
Shepody
Bay
Nappan
New Brunswick
Shepody
Maccan
Hebert
Chignecto
Bay
Nova Scotia
Intertidal flats
0
7km
Cumberland Basin

Map 13.

United States of America
Mexico
Gulf of Mexico
Teacapan -
Agua Brava
Pacific
Ocean

Map 14.

Teacapan

The Teacapan-Agua Brava-Marismas Nacionales lagoon system is positioned on the pacific coast of Mexico. It covers an area of 2000 km^2, marked by tidal channels, seasonal flood plains, lagoons and mangrove marshes. The territory of Teacapan and Agua Brava is largely covered with mangrove forests. The Marismas Nacionales consists of a series of parallel linear lagoons, which are divided by dunes. There are two outlets to the ocean. In the seventies an artificial opening was made in favour of the shrimp fishery, 43 km to the south of the natural inlet of the Agua Brava lagoon. The mean depth of the water system is 2 metres. The area is important for birds. The most frequent species are the heron and the pelican. Water pollution is not considered a substantial problem. The sewage discharge of the cities along the tributaries is the origin of coli type pollution, which however does not present major problems. Fisherman's trade is the most outstanding economic sector. The regional income depends on this activity for 65%. The high salt content of the soil and the regular floods obstruct agriculture. In the surrounding area extensive pasture is practised.

2.3. FUNCTIONS AND CONFLICTS

The following paragraph briefly describes the distinctive functions of the estuaries and their surroundings. The concept of function is used in a broad sense; it refers to the various uses by man of surface water, ground water, banks and shores, and surroundings of an estuary. Each function puts demands on the properties of the water system. As a result of these demands the functions of a water system and the activities on and in a water system can correspond, conflict or compete with one another. Table 4 presents an outline of the actual human functions for each estuary, at the same time indicating their relative importance.

This paragraph will deal with conflicting functions. Conflicts can be defined as an interference of functions. They arise whenever in some part of the estuary one of the uses harms another one substantially. For each estuary a cross table has been composed. Only those functions that lead to conflicts are represented in these tables. For economy of space, Capitals are used along the horizontal axis of the crosstables, which refer to the functions along the vertical axis. The conflicts, within the matrices, are represented by a number; for the serious conflicts these numbers relate to the corresponding text underneath the tables and can be found in the left margin of the page.

Tejo

Despite the severe pollution, the ecosystem of the estuary is considered to be in balance. The immediate problem is the bacteriological pollution by the city of Lisbon, which leaves much to be desired of the swimming water quality. The great number and intensity of functions will undoubtedly lead to complications in the near future.

Cumberland Basin

Until now, the isolated site of the Cumberland Basin prevented the emergence of any conflicts. The estuary is used rather extensively (see table 4). The construction of a tidal barrage will most certainly not take place before the year 2000. This dam, however, would lead to conflicts with fishery and the environment.

Table 4. Occurence of human functions

Estuary	Ebro	Huelva	Ave	Tejo	Loire	Venice	Solent	Eastern Scheldt	Western Scheldt	Cumberland	Teacapan
A. Waterbasin											
1. Shipping	0	3	0	3	3	3	3	2	3	0	0
maintenance	0	3	0	3	3	2	1	1	3	0	1
2. Fisheries	3	3	3	3	3	2	3	3	1	1	3
3. Aquaculture	3	2	0	3	0	3	0	1	0	0	0
4. Sand and gravel mining	0	0	3	3	2	0	2	1	2	1	1
5. Pipelines and cables	0	3	0	1	1	1	3	1	1	0	0
6. Water supply industry	3	3	3	2	3	2	3	0	1	0	0
7. Gas and oil extraction	0	0	0	0	0	0	0	0	0	0	0
8. Waste discharge	3	3	3	3	1	3	3	1	3	1	1
9. Tourism	3	2	2	2	1	3	3	2	2	1	1
10. Safety	0	0	0	0	0	3	1	3	0	1	0
11. Reclamation	0	1	3	2	3	2	2	0	0	3	0
B. Land use											
12. Ship building/port	0	3	3	3	3	3	3	1	2	0	0
13. Chemical industry	0	3	0	0	3	3	3	0	2	0	0
14. Other industry	2	3	3	3	3	2	2	1	2	1	0
15. Electricity supply	0	3	0	2	3	3	3	0	0	1	0
16. Roads	0	3	1	0	0	2	2	1	1	1	1
17. Residential areas	0	3	3	3	2	3	3	1	2	2	0
18. Tourism	2	3	2	3	1	3	3	2	2	2	1
19. Agriculture	3	3	1	1	3	2	2	3	3	3	2
C. Other functions		3 (salt)			3						

Absent	0	1	2	3	Occurs frequently

Crosstable Ebro (see Map 3)

	A.	B.	C.	D.	E.	F.	G.	H.
A. agriculture	\				1	2	3	4
B. aquaculture		\		5	6	7	8	
C. fishery			\		9	10	11	
D. tourism				\				
E. sewage discharge					\			12
F. industry						\		13
G. residential areas							\	14
H. nature								\

1 The most important agricultural activity in the Ebro delta is
2 the irrigated rice culture, followed by the cultivation of
3 vegetables and fruits. Due to the water supply to industry,
6 less water is available for this cultivation. The discharge
7 of waste matter by industries and cities upstream threaten
8 the quality of this water. Fishery and aquaculture are
9 affected as well. The reduction of the water supply and the
10 nutrients dissolved are harmful to the
11 fisheries and the shell fish culture.
Tourism, watersports in particular, is entangled in a
5 competition with the aquaculture over the use of the water.
Another spatial problem is the limitation of nature reserves by
4 the advancing agriculture. Already many nature areas
were cultivated in this way.
12 At this moment the interference of the draining with the
13 ecosystem is not yet alarming. However, a potential conflict
14 is at hand.

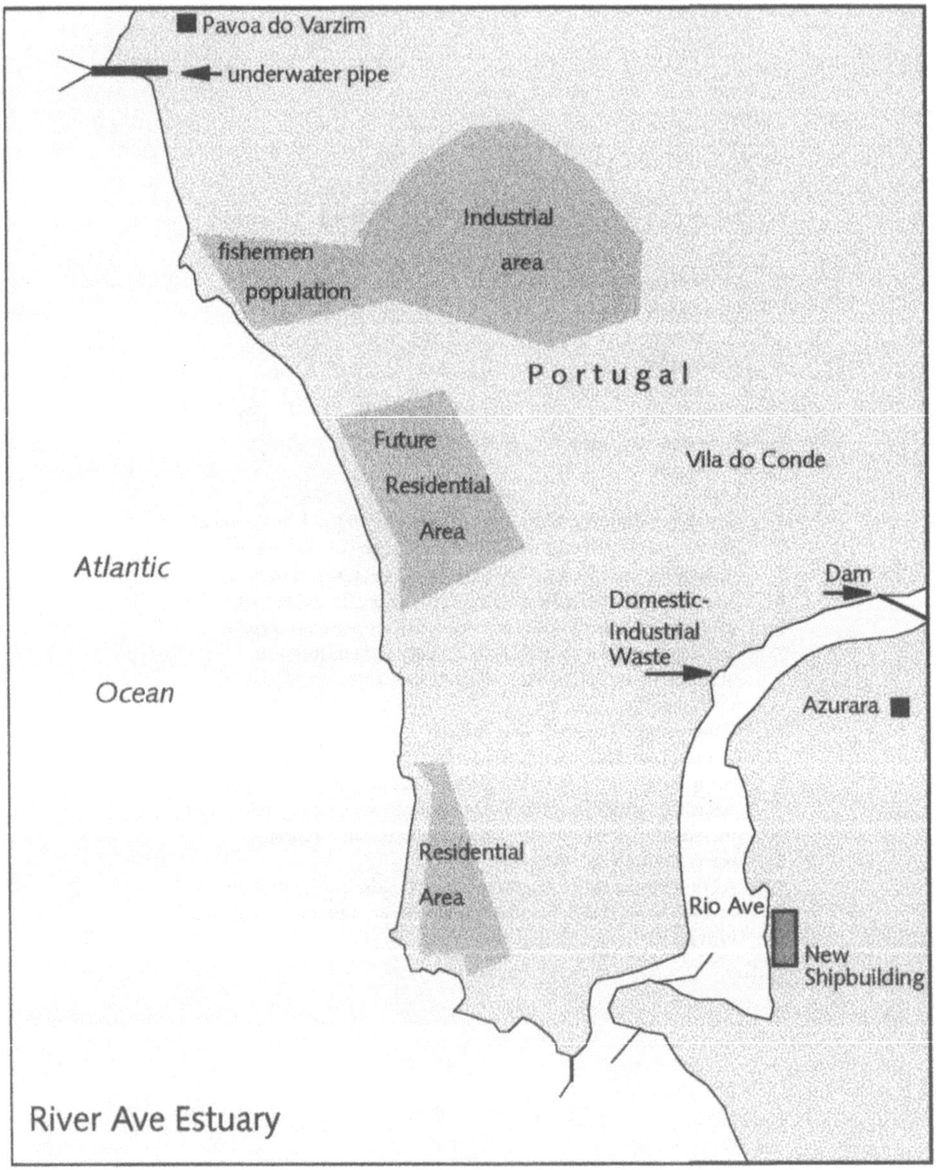
Pavoa do Varzim
underwater pipe
Industrial
area
fishermen
population
Portugal
Future
Residential
Area
Vila do Conde
Atlantic
Ocean
Dam
Domestic-
Industrial
Waste
Azurara
Residential
Area
Rio Ave
New
Shipbuilding
River Ave Estuary

Map 15.

Crosstable Huelva (see Map 4)

	A.	B.	C.	D.	E.	F.	G.	H.	I.
A. maintenance channels	\	1	2	3	4				5
B. fishery		\		6		7			
C. aquaculture			\	8		9			
D. sewage discharge				\	10		11		12
E. tourism					\	13			
F. industry and mining						\	14	15	16
G. residential areas							\	17	
H. agriculture								\	
I. nature									\

1 The maintenance of the main channels influence the entire area.
2 The water dynamics of the system and the discharges of solid wastes by
3 industries and mining necessitate intensive dredging. The
4 consequences are manifold. The dumping of spoil in the sea
5 causes considerable damage to the fishing of flatfish. By the
construction of the breakwater in order to counter silting
up, the sand movements changed substantially, which hinders
the aquaculture. Likewise beaches were partially washed away
by the altered dynamics, which diminished the shore
recreation terrain. Moreover, the underwater environment
suffered from these changes of the estuarine bottom.
6 Fishery and aquaculture are in conflict with the chemical
7 industry and mining because they worsen the quality of the water
8 by dumping residual products in the water and on land. This
9 has some disadvantageous consequences for watersports, shore
10 recreation and cities, since the environment, especially the
11 salt marsh, is affected.
12 Problems regarding the spatial use arise between the agriculture
13 on one hand and the industry and the city of Huelva on the
14 other. The prosperous economic development claims this area
16 increasingly.
15 It is expected that the conflicts will grow in number and
17 gravity.

Crosstable Ave

	A.	B.	C.	D.	E.	F.	G.
A. ship building	\	1					
B. industry		\				2	3
C. residential areas			\			4	5
D. agriculture				\		6	7
E. electricity supply					\		8
F. tourism						\	
G. nature							\

2 The discharge of effluents by the fish and textile industries
3 causes a decline in water quality. The agriculture and the
4 villages contribute to this pollution. The repercussions for
5 the environment are apparent. Also, tourism is retarded in
6 growth by the worsening swimming water quality (faecal coli)
7 on the bank of the estuary and the beaches.
Upstream dams for the generation of electricity hold much
8 water. The concentration of pollutants increases by the
diminished water supply. In the future, shipbuilding on
the north shore of the Ave mouth
1 will be enlarged. The discharge of unrefined oil of the fish
industry causes much trouble for this activity.

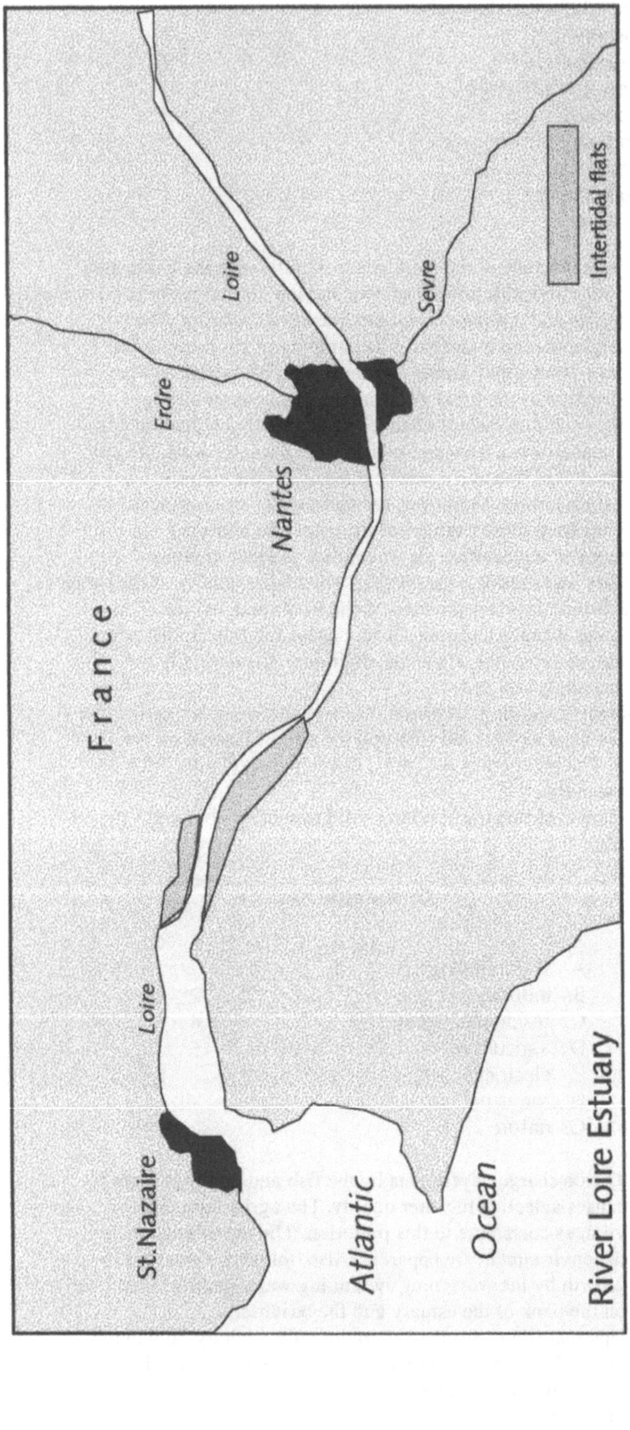
France
Loire
Erdre
Nantes
Sevre
Intertidal flats
Loire
St.Nazaire
Atlantic
Ocean
River Loire Estuary

Map 16.

Crosstable Loire

	A.	B.	C.	D.	E.	F.	G.	H.	I.	J.	K.
A. maintenance channel	\	1									2
B. fishery		\			3	4	5	6			
C. port-shipping			\						7	8	9
D. industry				\					10	11	12
E. residential areas					\				13	14	15
F. land reclamation						\					16
G. waste discharges							\		17	18	19
H. sand/gravel mining								\			
I. tourism									\		
J. agriculture										\	
K. nature											\

2 The wetlands on the Loire banks are under pressure. On the one
3 hand this is caused by the decrease of space by the extension of industrial areas, land reclamation and maintenance of channels and on the other hand by a high degree of pollution.

19 The deteriorating water quality is induced by industrial and urban drainages.
This development has negative results for watersports, shore recreation (swimming), agriculture (water supply) and fishery (number and quality of fish).

1 The fishery is hampered by the maintenance of the shipping lanes and the mining of sand, because these disturb the bottom. Moreover,
4 the surface area of water is reduced by the reclamation of land.
These conflicting situations are expected to be restricted by the implementation of management plans.

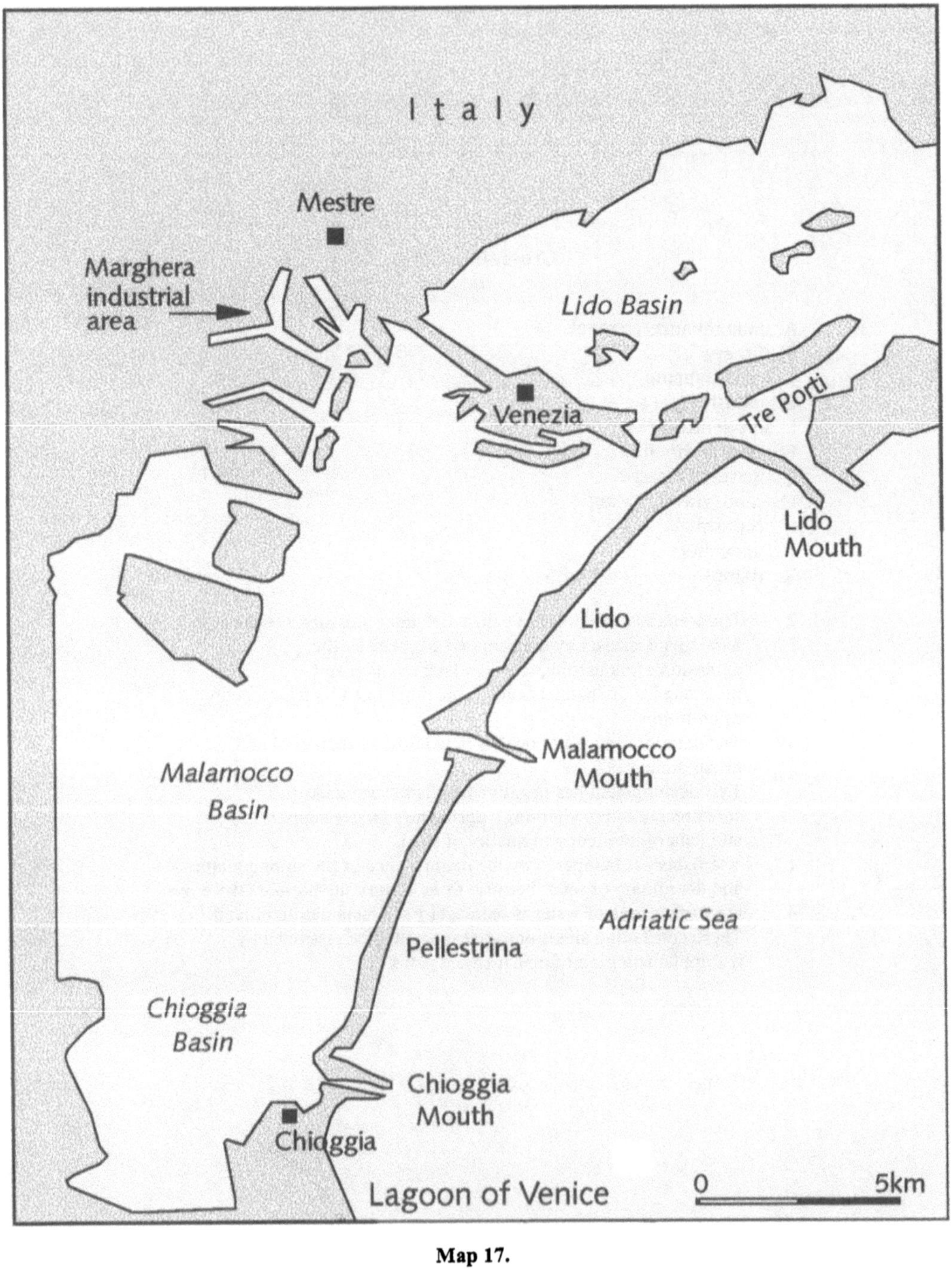
Italy
Mestre
Marghera
industrial
area
Lido Basin
Venezia
Tre Porti
Lido
Mouth
Lido
Malamocco
Mouth
Malamocco
Basin
Adriatic Sea
Pellestrina
Chioggia
Basin
Chioggia
Mouth
Chioggia
Lagoon of Venice
0
5km

Map 17.

Crosstable Lagoon of Venice

	A.	B.	C.	D.	E.	F.	G.	H.	I.	J.
A. fishery	\			1	2	3		4	5	
B. aquaculture		\		6	7	8	9	10	11	
C. tourism			\	12	13	14		15	16	
D. residential areas				\						17
E. industry					\					18
F. electricity supply						\				19
G. maintenance channels							\			20
H. agriculture								\		21
I. waste discharge									\	22
J. nature										\

1-8 In spite of the great number of conflicts at hand in the
lagoon, the explanation is straight forward. The water pollution
and eutrophication by the input of waste of industrial, agricultural
and domestic origin in particular, have a deleterious effect
on fishery, shellfish farming, water recreation and nature.
10 The residential areas of the lagoon islands, including
19 Venice, do not have any sewage purification plants, yet. The
effluents from industrial settlements are reasonably treated,
21 though a residue remains. By manuring, the agriculture
22 contributes to the eutrophic climate.
Two artificial channels of about 12 metres deep run from the
Lido and Malamocco mouth to the industrial (Port Marghere)
9 and the commercial harbour of Venice. The channel maintenance
20 has a disturbing effect on the underwater environment and the
shellfish farming. The algal bloom will get worse in the years to come.

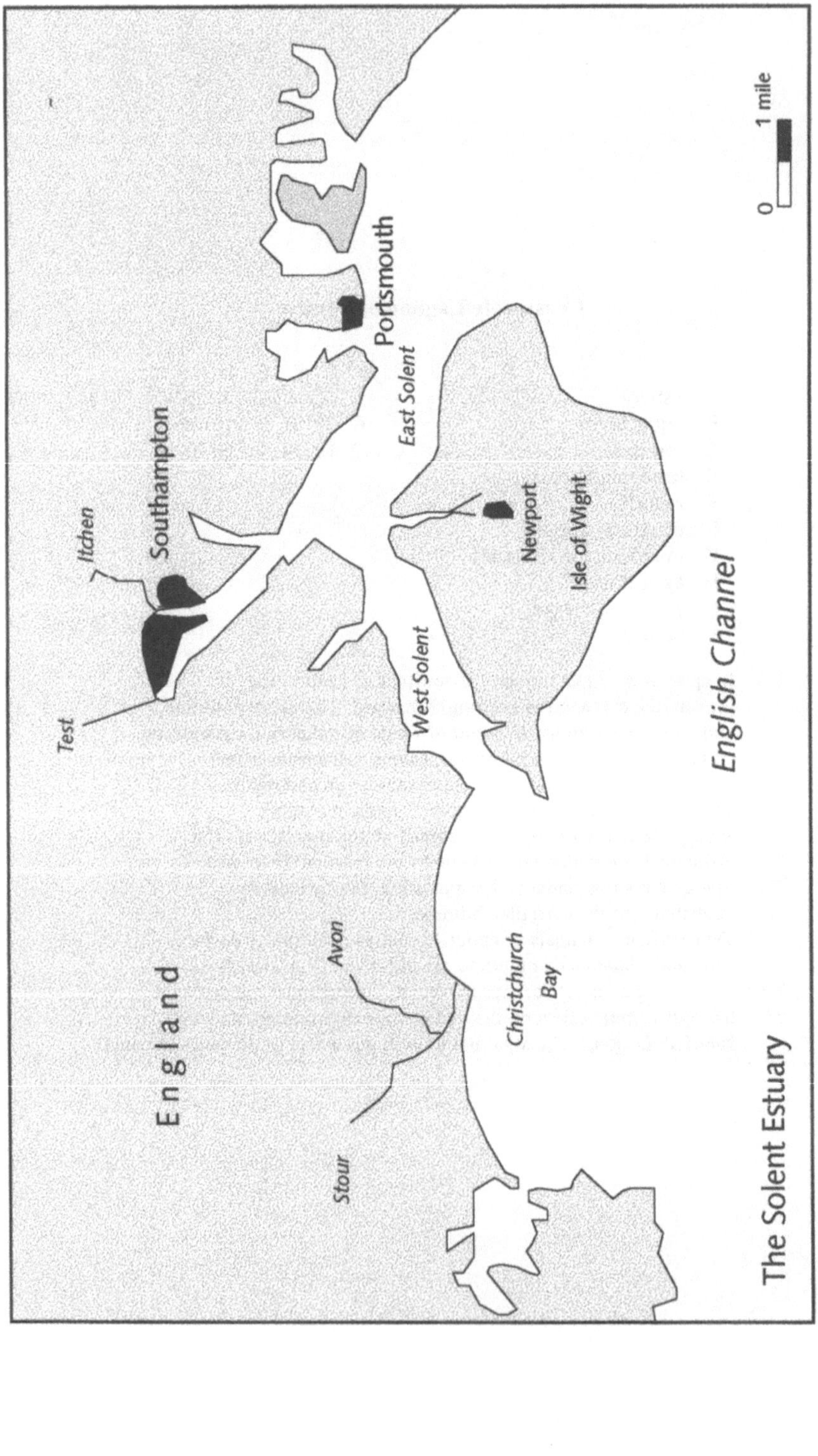
The Solent Estuary
England
Test
Itchen
Southampton
Portsmouth
East Solent
West Solent
Newport
Isle of Wight
English Channel
Avon
Stour
Christchurch Bay
0
1 mile

Map 18.

Crosstable Solent

	A.	B.	C.	D.	E.	F.	G.	H.	I.
A. shipping	\	1	2						
B. tourism		\							3
C. fishery			\					4	5
D. agriculture				\					6
E. sewage discharge					\				7
F. residential areas						\			8
G. industry							\		9
H. pipe lines/cables								\	
I. nature									\

6-9 The rather high algae production in the Southampton Water is attributed by agricultural run-off, treated sewage discharges and industrial inputs.
The environmental burden is not regarded as excessive. Flotsam on the
3 shores constitutes some minor nuisance. The sub-littoral region, and to
4 some extent the lower intertidal region, of Southampton Water has been
5 impacted by the effects of intensive clam dredging in recent years.
1 The concentration of yachts in the area is high resulting in occasional
2 difficulties for commercial shipping. Controls in the area include zoning,
4 licensing of discharges and restrictions on anchoring and fishing in certain zones.

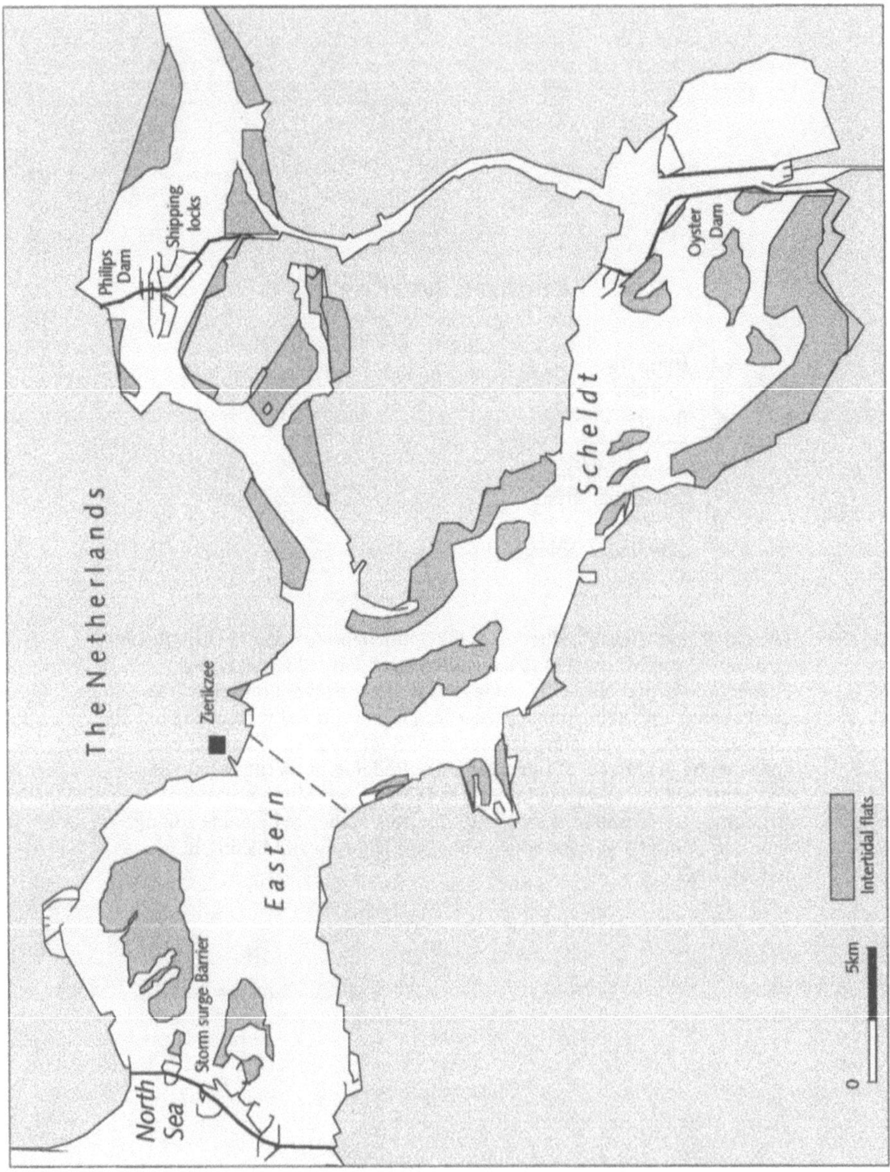
The Netherlands
Eastern
Scheldt
North Sea
Storm surge Barrier
Zierikzee
Phillips Dam
Shipping locks
Oyster Dam
Intertidal flats
0
5km

Map 19.

Crosstable Eastern Scheldt

	A.	B.	C.
A. fishery	\	1	2
B. tourism		\	3
C. nature			\

Many conflicts were eliminated by the framing of a policy plan for the Eastern Scheldt.

3 The main conflicts occur between tourism and nature. The shore- and watertourists upset the tranquillity in natural areas. Moreover, in the immediate surroundings of recreational concentrations some symptoms of water pollution appear by the draining of waste water, oil and preservation products.

2 The relation between fishery and nature is less conflictuous. This is a result of the limitation of capacity of the fishing on shellfish and the demands to proportion, weight and shape of the nets, which thus hardly disturb the underwater soil.

1 Fishery locally encounters an impact of recreation by the pollution with organotin used as an antifouler on boats, damage to bownets and the anchorage on shellfish parcels.

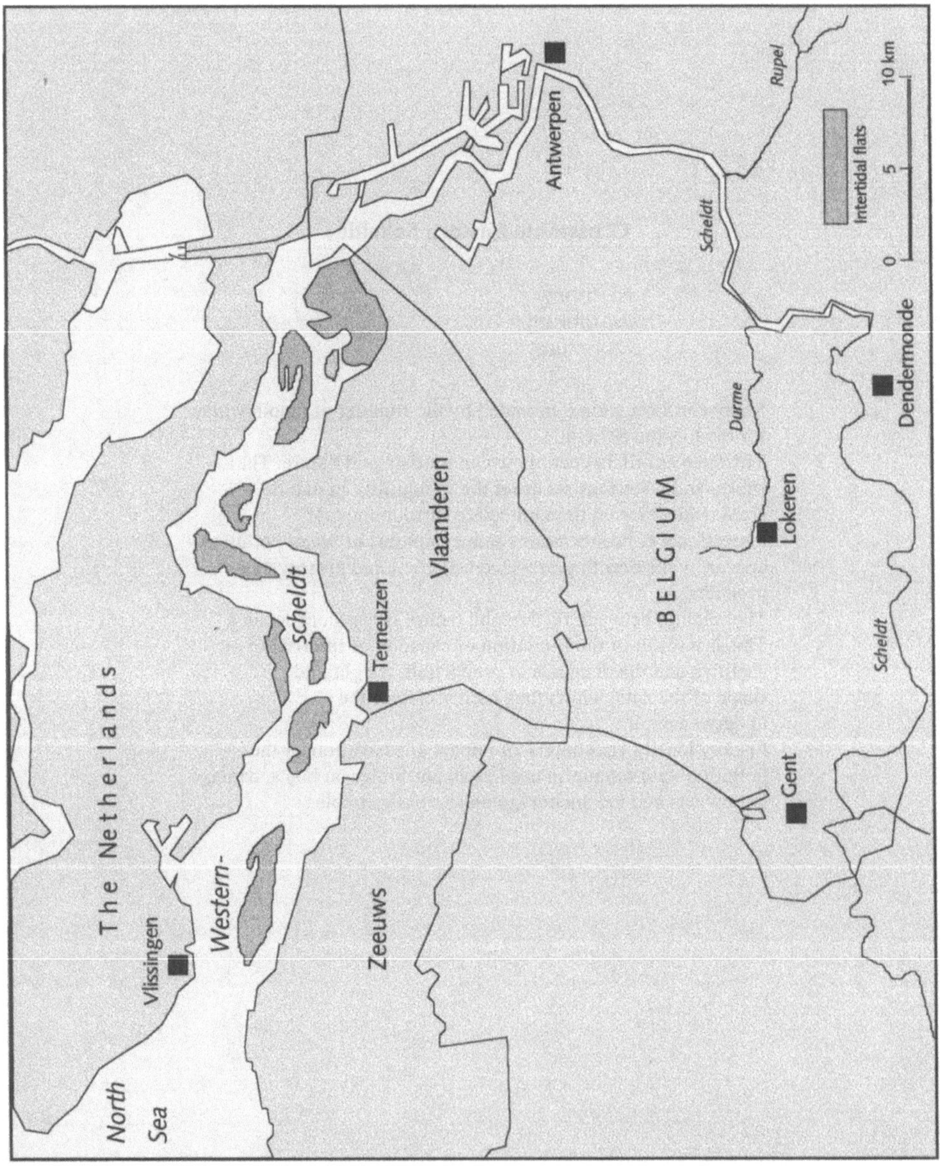
North Sea
Vlissingen
The Netherlands
Western-
scheldt
Zeeuws
Terneuzen
Vlaanderen
BELGIUM
Gent
Lokeren
Durme
Dendermonde
Scheldt
Scheldt
Antwerpen
Rupel
Intertidal flats
0
5
10 km

Map 20.

Crosstable Western Scheldt

	A.	B.	C.	D.	E.	F.	G.	H.	I.	J.
A. fishery	\					1	2	3		
B. tourism		\	4			5	6	7		8
C. shipping			\							
D. maintenance channels				\					9	10
E. sand/gravel mining					\					11
F. industry						\				12
G. residential areas							\			13
H. waste discharge								\		14
I. dikes									\	
J. nature										\

1 The Western Scheldt is heavily polluted by industrial and
2 domestic discharges especially from Belgium, which causes
3 conflicts with nature, fishery and tourism. For example, in
5 the eastern part of the river more diseased fish occurs than
6 in other estuaries and a hampered development of the salt
7 marsh vegetation is visible. The swimming water does not
always meet the standards as well. In addition to the
8 polluted water, the maintenance of the channels, the mining of sand
10 and tourism are a burden for the environment too.
14 Dumping and dredging of spoil and mining of sand distort the
river bed and the natural accretion of the intertidal
areas. More over these activities threaten the water
9 retaining function of the dikes contributing to undermining.
The mooring along and trading on isolated territories by
tourists and worm diggers disturbs birds, especially in the breeding season.
4 Watersports, sea shipping and inland navigation make use of
the same waterway in large parts of the Western Scheldt.
Dangerous situations occur regularly by differences in
speed, proportion and course of the boats.

Beside these actual conflicts, many potential conflicts are expected to develop. In order to control this development a policy plan has been conceived recently, which takes all conflicts into account.

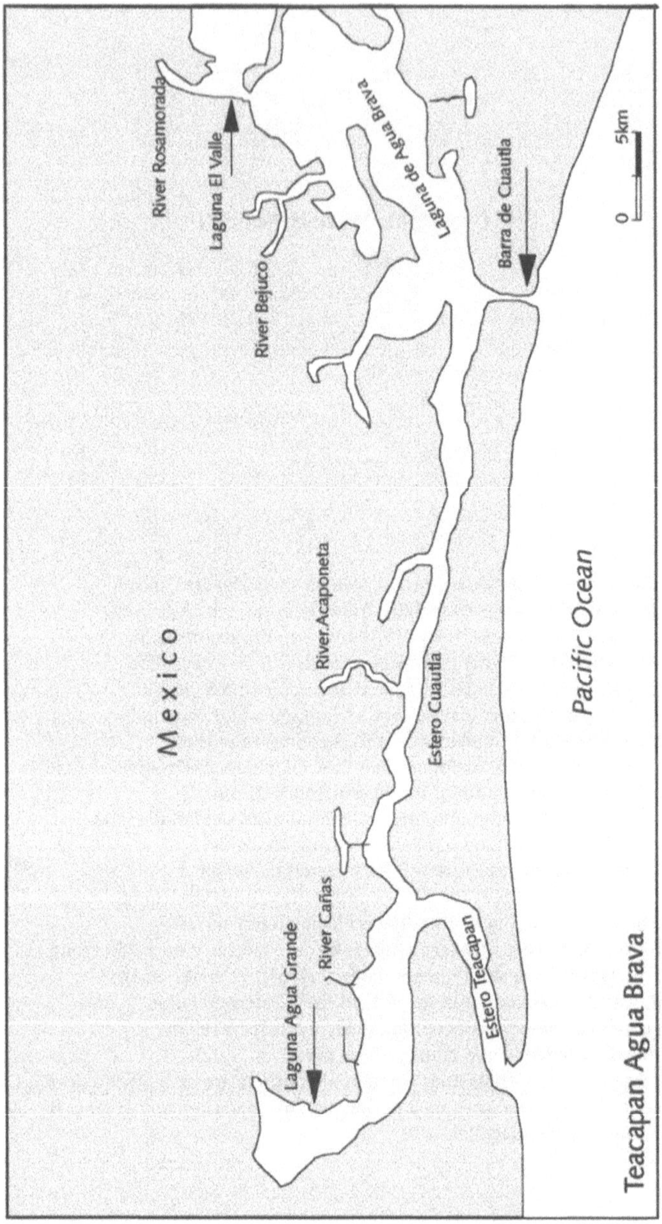
Mexico
River Rosamorada
Laguna El Valle
River Bejuco
Laguna de Agua Brava
Barra de Cuautla
0 5km
River Acaponeta
Estero Cuautla
Pacific Ocean
River Cañas
Laguna Agua Grande
Estero Teacapan
Teacapan Agua Brava

Map 21.

Crosstable Teacapan

	A.	B.	C.	D.	E.	F.	G.
A. fishery	\	1		2	3	4	
B. agriculture		\					5
C. accessibility			\				6
D. artificial inlet				\			7
E. residential areas					\		8
F. sewage discharge						\	9
G. nature							\

The hydrodynamics of the lagoon changed completely after cutting an artificial channel. The reinforced tidal movements produced higher salinity figures, benefiting shrimp fisheries, but hampering the catch of other fish.
Likewise, road construction has changed these dynamics, which caused the disappearance of several mangrove swamps.
The eutrophication, induced by the agriculture and villages upstream, did harm fishery and the environment.
In the near future the intensive aquaculture will probably increase, which has negative consequences for water quality, larval populations and the equilibrium of the ecosystem.

2.4. WATER QUALITY

Water is a basic natural resource. The quality of the water is essential for the functioning of estuaries. In this paragraph a tentative classification on the water quality will be presented.

One was asked to submit data for three sites, namely the head, the intermediate and seaward location of the estuary. At first sight, the lack of data was striking. It appears that sampling of water quality is often carried out for a limited number of parameters only. Furthermore, some of the measurements are outdated and the quality of the data given is not always correct, either. Because of outdated methods higher and lower concentration values were given in some cases. Especially the tracing of organic micropollution poses problems for lower concentrations. Nevertheless, it is essential to measure the extent of pollution of water, bottom, flora and fauna, in order to generate efficient planning and control.

Comparison of the figures is subject to uncertainties. Difficulties that arise in this area of concern stem from the sliding figures of salinity in estuaries. Chemical reactions change concentrations of pollutants, among others influenced by the increasing chloride content towards the sea. Salinity classification of brackish water is still being discussed internationally, although provisional ones have been proposed.

In order to allow for some comparisons to be made, standards were set up based on several existing classifications. Table 5 presents the tentative standards. A threefold division has been made, because the concentration of pollutants diminishes in the direction of the sea, induced by biological, chemical and physical processes. The classification distinguishes three salinities, viz. 0-6, 6-10 and more than 10 0/00 Cl^- per litre, named I, II and III respectively. The columns beneath these classes present the values with which a water

Table 5. Classification waterquality, standards

		Classification		
		I	II	III
Salinity 0/00 Cl^-/l (variable)		0-6	6-10	>10
General parameters				
Dissolved oxygen	(mg/l)	5	8.4	8.7
Chlorophyll	(μg/l)	<100		<10
Ph			6.5-9	
Tot. coli	(MPN/ml)	0	0	0
BOD 5	(mg/l)			<5
Phosphate tot. P	(mg/l)	0.15	0.09	0.03
Phosphate ortho. P	(mg/l)	0.10	0.06	0.017
Nitrogen tot. N	(mg/l)	0.61	0.45	0.21
NO3 + NO2 tot. N	(mg/l)	0.52	0.39	0.19
Inorganic parameters				
Cd tot.	(μg/l)	0.28	0.21	0.10
Hg tot.	(μg/l)	0.13	0.09	0.04
Cu tot.	(μg/l)	5.7	4.5	2.7
Pb tot.	(μg/l)	7.4	5.6	2.8
Cr tot.	(μg/l)	14.6	11.4	6.2
Organic parameters and oil				
PAK (Σ)	(ng/l)	25	13	1
Phenols	(μg/l)	0.2	0.1	0
Oil	(μg/l)	140	95	30
PCB	(ng/l)	0	0	0

Table 6. A tentative classification of pollution degree

Estuary	Number of minusses (-)	Number of plusses (+)	Quotient of minus and plus (-/+)	Classification
Western Scheldt	52	15	3.6	severely polluted
Venice	16	5	3.2	
Tejo	23	8	2.9	
Loire	29	11	2.6	
Ave	11	7	1.6	moderately polluted
Huelva	11	8	1.4	
Ebro	9	8	1.1	
Solent	9	9	1	
Eastern Scheldt	10	31	0.3	hardly polluted
Cumberland	3	13	0.2	

sample, of a certain salinity, should comply. The measured values of the water quality parameters are compared with the standards. Minus (-) indicates a bad result for a particular parameter (more than the standard value), plus (+) indicates a favourable score (less than the standard value). Furthermore, a double minus (– –) stands for extreme exceeding of the standard value and a double plus (++) for an excellent score. The assessment of this last result (++ and – –) has been conceived as follows: the highest concentration of a pollutant is taken as a maximum. A double minus has been assigned if a concentration ranges between 50 and 100% of this maximum. A very good score produces the opposite result. Table 6 presents the final outcome of this analysis.

Almost all estuaries are polluted to a certain extent. The Eastern Scheldt and the Cumberland Basin present positive exceptions. This is hardly amazing, in view of the kind of conflicts in these estuaries. Drawing definite conclusions based on this table is a hazardous operation due to the lack of data and the gravity of pollution. For indicative reasons only, table 6 presents the quotient of the total of minuses and pluses derived from table 6. Based on this quotient, a tentative classification of the estuaries is framed. The Teacapan is left out because only five parameters have been investigated. It should not be forgotten that this classification does not take into account the seriousness of pollution. Again, Eastern Scheldt and Cumberland Basin turn out to be the least polluted estuaries. Most heavily polluted are Western Scheldt, Loire, Lagoon of Venice and Tejo. The Huelva, Ave, Teacapan, Ebro and Solent are moderately polluted.

2.5. COMPARISON OF CONFLICTS

Table 7 compares the score of conflicts and functions in and alongside the estuaries. The conclusion is obvious. The number of conflicts outnumbers the functions. Seven water systems exceed the set standard of eight functions and conflicts. Two estuaries score below this standard. The Teacapan in Mexico presents an exception. Fishery and agriculture are the sole intensive users of this area. On the other hand, nine conflicts play a role. This is caused by the sewage discharges of cities upstream and the construction of an artificial inlet, which were not incorporated in the table of functions (table 4). The Solent performs positively as the number of conflicts is considerably lower than the number of functions. The reason for this discrepancy will be dealt with below.

Table 7. Functions and conflicts in number

Estuary	Number of Functions	Number of Conflicts	Functions/ Conflicts	Functions - Conflicts
Ebro	8	14	0.6	-6
Huelva	17	17	1	0
Ave	10	8	1.3	2
Tejo	14	?	–	–
Loire	13	19	0.7	-6
Venice	17	22	0.8	-5
Solent	16	9	1.8	7
Eastern Scheldt	6	3	2	3
Western Scheldt	11	14	0.8	-3
Cumberland	4	0	–	4
Teacapan	2	9	0.2	-7

Functions: numbers of functions with an intensity of 2 and 3 in Table 4 are added.

In conclusion, if an estuary serves many purposes, many conflicts are at stake. Almost all conflicts can be reduced to two components, being pollution on the one hand and the use of land or water on the other. Only proper management strategies may control these problems adequately.

It will be noted that to some degree the presented tables are derived from subjective judgment. The interviewees made estimations about the intensity of a function and the presence and seriousness of a conflict. Yet, the pollution figures are objective. In almost all estuaries pollution has a large impact. In most cases, fishery, aquaculture, tourism and nature encounter difficulties caused by the deteriorating water quality originating from sewage discharges of industries, residential areas and agriculture. Within every description of the estuaries, one or several elements of these problems emerged.

2.6. MANAGEMENT PLANS

What about the quality of management of the estuaries in question? What are the elements that constitute the plans? These are some of the questions that will be answered in this part. First of all, the contents of the plans will be described briefly per estuary and masterplans will be distinguished from partial plans. The masterplan is an integral plan, encompassing the entire estuary and covering all sectors involved. Partial plans are restricted to a specific topic in order to deal with a part or a separate sector of the estuary. This distinction of masterplan and partial plans numbered 1, 2, and 3 is reflected by tables 8 - 13. In these tables, the elements of the plans will be put forward. These refer to: regulation, objectives, descriptions, research, measures, initiators and participants. Finally, the different management strategies will be compared. Little can be said about the effectiveness of the plans, as most were initiated rather recently.

The Ebro and Tejo will not be considered in this context as plans have not been constructed for these estuaries, as yet. The interviewees of these water systems did remark that there is an urgent need for planning, caused by the large number of conflicts. They report the initiation of planning.

This paragraph outlines a description and evolution of the plans that are appropriate to the estuaries involved. The text will also deal with the role of the authorities responsible for the preparation and implementation of the plans.

Huelva

At the beginning of the eighties, the causes, results and gravity of the pollution of the Huelva were examined extensively. As a result of this investigation, a masterplan was developed for the Odiel and Tinto areas in order to sanitate industrial discharges. Its objective was to set bounds to the draining of solid and liquid wastes. For this purpose, a water purification plant and a properly controlled dumping ground were built. In addition, a monitoring programme for the control of water quality was drawn up. The national directorate-general for the environment and the Andalusian Board bear the final responsibility for the correct execution of the plan. Many tasks were delegated to lower administrative councils and research institutes.

At the moment, it is not possible to give a clear indication of the effects of these plans.

Ave

The CGIBHRA (Commission for integrated management of the Ave river basin) was founded about 1985 in order to keep the developments in the area under control. A masterplan was schemed by the commission in order to solve the pollution problems of the area. This plan was backed by the national government and was set for a five year term. During this period the degree and sources of pollution of the river were examined, and a beginning was made with an action plan for the taking of concrete measures for the sanitation of the water quality. The Ave estuary was not included within this plan, but the upstream activities were coordinated with the city councils along the estuary.

The partial programmes cover the control of new industrial settlements, the construction of water purification plants and regulation and processing of the waterflow respectively. All the plans are part of the masterplan.

Loire

On September 16, 1970, the French cabinet adopted the SDAAM (Schema d'Amenagement de l'Aire Metropolitaine de Nantes St. Nazaire) masterplan. This long term programme is binding for lower government authorities and aims to bring about a controlled economic and physical expansion of the region. The plan is implemented by the OREAM (l'Organisation d'Etudes d'Amenagement de l'Aire Metropolitaine). Local governments and institutions concerned, such as the chamber of commerce and agriculture as well as port authorities are represented in the coordinating committee. Also, the OREAM has the disposal of a research department in which economists, sociologists, geographers and town-planners take part. The organisation focusses on the enlargement of the agglomeration centre function and the stimulation of industrial and harbour activities. Some space has been allowed for the development of agriculture and recreation, as well. Local authorities take care of the detailing of the principles. They decide on matters such as land allocation and the site of infrastructural projects and expansion areas.

In 1980, the Port Autonome de Nantes, a port authority, founded the Comité Scientifique pour l'Environment de la Loire, which conducts multidisciplinary research on the state of the estuary environment. In 1984, the committee gave account of their activities. Besides giving a description of the hydraulic and sedimentary aspects water quality, flora and fauna, recommendations were made for sound control of the salt marshes on the north shore of the estuary. The recommendations relate to the organisation of the estuary, nature conservation and the improvement of water quality control. In 1984, following this report, the APEEL (l'Association pour la protection de l'Environment de l'Estuaria de la Loire) was founded, in which representatives of local authorities, public utilities, consumers, and nature

Table 8. Regulation in Master plans (MP) or Partial plans (P)

	Estuary																	
	Huelva	Ave				Loire	Venice			Solent				Eastern Scheldt				Western Scheldt
Plan	P1	MP	P1	P2	P3	MP	P1	P2	P3	MP	P1	P2	P3	MP	P1	P2	P3	MP
The plan regulates																		
1. Physical planning																		
a. land use		Y								Y	Y			Y	Y	Y		Y
b. spatial organisation		Y				Y				Y	Y			Y	Y	Y		Y
2. Water management																		
a. quantitative	Y	Y	Y				Y			Y		Y						
b. qualitative	Y	Y	Y	Y	Y			Y	Y	Y							Y	Y
3. Nature conservation/restoration	Y	M			Y			Y		Y	Y	Y		Y	Y	Y		Y
4. Landscape conservation	Y	M			M					Y	Y	Y		Y	Y	Y		Y
5. Planning economic sectors	Y					Y												

Y = apparent in plan
M = moderately apparent in plan
= not apparent in plan

conservationists take part. The APEEL maintains a control network for water quality, and produces research proposals - whether asked or not - on the environment and in particular on water quality. Special attention is paid to the limitation of pollution effects, as well. They also advise on the desired future structure and use of the estuary and inform the public on the environment.

The effect of the foregoing plans on the degree of pollution is not measurable, as yet.

Recently, two substantial improvements in the planning of the river and estuary have been produced. Firstly, an 'Interministerial Committee of Land Management' has been established producing an exhaustive 'Relevé de Décisions' which deals (in part) with the estuary. Secondly, a designated three dimensional mathematical model has been processed by ACEL (Association Communautaire de l'Estuarie de la Loire), which associates local authorities in coordination with the state regional level, in a so-called 'Comité de Pilotage.'

Lagoon of Venice

In 1973 already, the Italian government initiated a masterplan for the Venetian region. The regional authorities were entrusted with the scheming of the plan, entitled Plano di Area della Lagune Veneta. Because this plan has not been completed yet, it will not be taken into account in this analysis. The plan is characterised by the integral approach of all human activities in the area and allocation of part of the lagoon to natural reserve.

The partial plans that are in effect at the moment in this area, will be incorporated in the masterplan.

The first subprogramme intends to control the water level of the lagoon. For this purpose, engineering works are planned for the closure of the portsmouth with a mixed system of fixed and mobile flood gates, and for the consolidation of littoral barriers and natural shore lines.

The second subprogramme relates to the improvement of the water quality. In 1977, this plan was accepted by the Regione Veneto, directed to implement the collection and treatment of waste waters of industrial and urban origin. The goal is to reduce the pollutant concentrations below standards that are set by national laws. Two problems are not resolved yet: the collection and treatment of sewage in Venice and diffused sources of pollution due to agriculture. These items will be considered in the general plan.

The last subplan deals with the water quality with regard to the shellfish culture and swimming water. Standards and regulations are set by the national government by law, and equal to other Italian waterbodies. Until now, the laws in question have not produced sufficiently tangible measures on local levels. It is expected that the situation will improve after completion of the masterplan, which needs to be agreed upon by all authorities concerned.

Solent

The masterplan prepared by the Hampshire County Council covers the greater part of the Solent and concerns physical planning and environmental quality. It is a non-statutory plan based on cooperation between the authorities involved. In Great Britain, the situation has emerged wherewith the National Rivers Authority is responsible for the policy with regard to pollution. This set of problems lies beyond the jurisdiction of the county councils. Nevertheless, the masterplan does take qualitative water management into account.

Subprogramme number one, the Portsmouth Harbour Local Plan, only affects the activities and land uses around the total harbour area of Portsmouth. The river Hamble Local Plan is part of this first section of subprogrammes as well. It controls the mooring numbers in Humble river and activities at boat yards and other waterside sites.

Table 9. Starting points and policy objectives

	Estuary																	
	Huelva	Ave				Loire	Venice			Solent				Eastern Scheldt				Western Scheldt
Plan	P1	MP	P1	P2	P3	MP	P1	P2	P3	MP	P1	P2	P3	MP	P1	P2	P3	MP
1. Reducing pollution		Y	Y	Y	Y					A								Y
a. status quo																		
b. best technical means								Y									Y	
c. best practical means	Y																	
2. Polluter pays	Y							Y										Y
3. Quality objectives/standards				Y	Y												Y	Y
a. drinking water		Y																
b. fisheries	Y	Y							Y								Y	Y
c. swimming water	Y	Y							Y								Y	Y
4. Ecological conflict solving by																		
a. multifunctional zoning										Y								Y
b. monofunctional zoning										Y								
5. Implementation of functions																		
a. one function																		
b. several functions						Y				Y								
6. Human activities																		
a. reduction																		
b. continue		Y	Y							Y								Y
c. increase																		
7. Integration																		
a. environmental policies														Y				
b. other policy areas										Y				Y				Y

A = Responsibility of National Rivers Authority

Y = apparent in plan

The second subplan contains the local plans of the coastal district councils. These include coastal sites but there are few coastal policies. The Hampshire County Council owns much land on the coastline and has prepared management plans for many of these sites. They include the Lymington Coastal Management Plan, the Hook with Warsash Local Nature Reserve Plan and the Chilling Coast Management Plan. In Great Britain, a rather unique situation has occurred in regard to waterpollution. This is regulated by a number of National River Authorities and lies beyond the jurisdiction of the county councils.

The former head of the Hampshire County Council planning department stresses the fact that coordination between agencies and cooperation between adjacent authorities is necessary. Only in this way an integrated policy is possible wherewith local plans are incorporated into a masterplan. He points out the importance of following a simple policy. This means: policies easy to understand and implement, clear and not too high set objectives.

Eastern Scheldt

In 1977 the province of Zeeland established a steering group Eastern Scheldt. This group embodies representatives of the national, provincial, and local authorities. It advises on matters of the Eastern Scheldt. It also organises and manages the basin. In 1982, after public hearing procedures, the policy plan for the Eastern Scheldt was accepted by the steering group. The plan confines itself to main policies of organisation and management of the area. It contains the following elements:

- The policy to be pursued with regard to the main functions.
- A planning outline which reflects the projects and spatial developments considered.
- A broad description of the desired management.
- An outline of research to be carried out in the 1980's for the evaluation and possible review and supplements of the policy plan.

This plan is not legally binding. One has opted for a declaration of intent concerning the coordination of the policy of the Eastern Scheldt. The plan is the starting point of the policy to be conducted and is embodied in the plans of national, regional and local authorities. Therefore, this policy plan could be seen as the masterplan. The partial plans in table 9 relate to the plans of the authorities concerned. The most important ones are, on national level, the designation of the Eastern Scheldt to Protected Nature Monument (1988) and the Pollution of Surface Waters Act. On a provincial level the Regional Plan Zeeland and the Overall Vision Delta Waters Note are of main concern. These plans regulate the spatial planning of the area in general terms. The Land Use Plans of local authorities provide for the details.

Water Authorities control the quantity and quality of water in compliance with the policy plan. The other partial plans too subscribe and reinforce the main policies of the masterplan.

As far as the masterplan is concerned, the main goal to be achieved is formulated as follows:

- The conservation and reinforcing as far as possible of the present natural values, considering the basic conditions for the socio-economic functioning of the area, among which primarily fishery.

This main purpose may not violate the prime goal of the Delta Works, i.e. the safety of the people. Then the following hierarchy of priorities has been derived: 1. Nature, 2. Fishery and 3. Recreation, shipping, etc. In the policy plan, this hierarchy is always apparent.

In 1992 the policy plan for the Eastern Scheldt was evaluated. Extensive studies show that the main features of the policy plan are still valid. One of the most important changes in policy is a greater emphasis on the development of natural values for example by restoring the links between the nature areas behind the dikes and the Eastern Scheldt. The development of fisheries and recreational activities will be between the lines drawn by the first policy plan.

Western Scheldt

The project structure with regard to the Western Scheldt has many similarities to that of the Eastern Scheldt. In the beginning of 1986, the Government, the Province of Zeeland and the municipalities concerned, decided to draw up a plan for the management of the Western Scheldt. For this purpose, a governmental control board and a core group were installed. This board can be compared to the steering group Eastern Scheldt, in that all authorities who have a say in the matter are represented. The core group has an advising task and consist of provincial and national experts. In view of the international context, Belgium is represented by one observer in each of the groups. In order to found the policy plan, the core group has conducted a number of studies. After public participation on a concept of the plan, the final version was published in 1991 (Bestuurlijk klankbordforum, 1991).

As with the Eastern Scheldt, the authorities commit themselves by signing an agreement.

> The main goal is to create a situation in which the environmental functions of the estuary can be maintained and restored and potential natural features developed, whilst preserving the role of navigation in the area and its potential for development (including the associated seaport and industrial activities), the aim being that this should also afford a satisfactory basis for developing fishing and recreation in the area.

The policy plan has taken into account the national plans already effective or in preparation. Other partial plans adapt their aims and further elaboration of this policy plan.

Conclusively, it can be stated that this paragraph highlights the diversity in planning and management of the estuaries concerned. As such, it is quite useful to gain insight in the variety of planning approaches. However, this does not necessarily produce a planning rationale that might be helpful in other situations. The following will attempt more specifically to develop an adequate planning classification. This will be based on the apparent degree of difficulty in approaching water systems.

2.7. EMERGING LEVELS OF PLANNING

Saeijs hypothesized that management strategies will differ according to levels of pollution. This study cannot fully deal with this matter as only the details of a restricted number of cases, and in varying degree of specificity, are available. There has been a considerable 'non response' in relation to the potentially addressed population. This was to be expected regarding the relative new issues at stake and the varying degree of water management skills all over the world.

As such, this study is about emerging considerations with water systems. The situation is characterized by growing research and planning efforts, within different stages of development, and the pattern of which is rather scattered. Instead of a comprehensive description of planning developments of water systems, this study basically seeks to identify problem situations and the way they could be approached.

The matter at stake might be expressed as the search for adequate approaches towards estuarine management; the reason for management and the stages it goes through. As such, it concerns the eventual shaping of planning profiles, which may be applied to water systems.

From the evidence gained from this study it is assumed that differences in water management may be related to the amount of conflicts demonstrated by water systems. A tentative classification might include three types of problem situations:

Type A Problem situations which are not subject to management or planning, because neither the need nor the opportunity have yet arisen.

Type B Problem situations with a certain amount of conflicts between the existing functions, which, however, are subject to more or less adequate planning approaches.

Type C Problem situations which are perceived widely as such, but, because of their severity and complexity have not yet brought about any successful planning effort.

What reasoning results from this classification? One might postulate two straightforward rationales towards planning development of estuaries. The first takes a down to earth or even negative point of view: the more conflicts one faces the more difficult it is to develop some planning approach, i.e. planning will only develop under rather simple circumstances. Although this may be true for a number of actual planning situations, it is also apparent that the more conflicts one is facing, the more a call for intervention may be heard. Therefore, the second rationale is normative, stating that the more conflicts one is facing, the more the need for planning emerges.

Firstly, one might expect in general far more estuarine water systems not or hardly being managed than emerges from this study, simply because the need or the possibility has not been apparent until now ('category A' water systems). We will return to this subject below.

Secondly, in what way may levels of estuary planning be distinguished according to some actual degree of complexity of the estuary problems?

Figure 4. Mud flats in the Solent provide food for bottom fauna and birds.

The issue may be approached by considering the relationship between functions and conflicts (referred to as f and c) in an estuary.

One may express this relationship in two ways (1) by the f/c ratio, or (2) by the f-c difference (table 7). Both happen to produce a similar twofold classification; hence, the water systems may be characterized on the one hand by having a greater than 1 value or a positive value - more functions than conflicts ('category B' water systems), and on the other hand by having a lesser than 1 value or a negative value - more conflicts than functions ('category C' water systems).

From the tables it emerges that planning is more likely to develop fully in category B (a 1+ or positive f-c value) than in category C cases (of a 1- or negative f-c value). All 1+ ratios, with the exception of Cumberland Basin, belong to this category, being the Solent, Eastern Scheldt and Ave. Lower f/c values show, with the exception of the Western Scheldt, planning situations in statu nascendi, or none at all.

The rationale might be that the lesser the extent of problems (conflicts) involved, the easier it will be to develop a comprehensive strategy and the corresponding implementation instruments. Evidently, along this line of view, the reverse will lead to a more difficult situation to control.

However, in each of both cases, an exception does blur the attractive simplicity of this inference.

The Cumberland Basin has an extreme high f/c ratio although there is no evidence from the questionnaire that a developed planning situation is as yet present. However, this will also be the case in many of the non response cases, as one might still expect the existence of many coastal water systems characterized by some functions but no evident conflicts as they are away from intensive human activities ('category A' water systems).

Figure 5. The Western Scheldt is a main shipping route to Antwerp, Ghent, Terneuzen and Vlissingen. Dredging has a disturbing effect on the ecological system.

So negligence of planning may be considered normal where no conflicts are present at all. In other words, at least some conflicts are a prerequisite to trigger planning efforts on the one hand, although, on the other hand, they should not be so abundant in number that development of planning is going to be an extremely difficult exercise. As such the Cumberland Basin does not have a high propensity towards planning — for the time being.

Compared to the Cumberland Basin, the Western Scheldt case is an exception at the other side of the range. The conflicts even outnumber the functions, yet the planning situation has matured to a level which goes beyond the expectations emerging from this study. What explanation is available that might be satisfactory enough as not to really undermine the rationale that too many conflicts also counteract planning development?

Unlike the evidence gained from the other cases, the Western Scheldt planning is not standing on its own, and it is hard to believe that it might have developed as a single activity out of nowhere. Actually, planning development of water systems previously started as a part of the Deltaplan activities. Following from these experiences the Western Scheldt planning effort was initiated. As it made direct use of the highly developed Eastern Scheldt planning skills, the vast problems of the Western Scheldt estuary could be traced down and put into an effective procedure. Lacking this experience it might have proved to be unsurmountable or at least be extremely difficult to overcome on a solitary base.

Hence, the Western Scheldt case is rather unique considering the developments up until now in general. One might view it as belonging to a class of its own ('category C' water systems), representing the planning of an intensively used and maltreated water system. Compared to the other evidence gained from this study the number of conflicts in relation to the functions would normally have resisted adequate planning development.

2.8. SOME PRELIMINARY CONCLUSIONS

Even the best developed planning structures show omissions. Generally, the missing parts are found in the areas of the formulation of objectives and (evidently related) the specification of measures to be taken. Inventories and research support are available in well developed planning systems, but apparently far less present in the emerging planning environments. Obviously, the development of planning systems is very much dependent upon data gathering operations and related research projects. In this context it should also be noted that economic planning usually lacks in relation to the water systems planning. In the long run it will be unavoidable to relate the two more closely, as the interdependencies are obvious.

Preparing the plans, research into these matters is a first and foremost step towards understanding many of the mechanisms involved, especially when category C water systems are at stake.

This study shows that knowledge of the number of functions and conflicts of estuaries may provide relevant information concerning the severity of the problems as well as the propensity towards planning of water systems.

Therefore, as it is likely that investigations of this rather straightforward nature will not demand highly specialized skills, it is important to stimulate, at least, an investigation of functions and conflicts of estuaries in all suspect areas of the world. Such an investigation may provide the basis for a mega strategy concerning planning development. Priorities can then be set in order to improve the situation.

If the comfort of any choice is available, planning developments should initially be encouraged in those water systems, which reveal a proper ratio of functions and conflicts, i.e. water systems where functions outnumber the amount of actual conflicts. Planning is likely to be successful in these conditions. Furthermore, it is highly important to have the

Table 10. Inventories and Descriptions

	Estuary																	
	Huelva	Ave				Loire	Venice			Solent				Eastern Scheldt				Western Scheldt
Plan	P1	MP	P1	P2	P3	MP	P1	P2	P3	MP	P1	P2	P3	MP	P1	P2	P3	MP
1. Environment:																		
a. characteristics		Y				Y				Y	Y		Y	Y	Y			Y
b. value		Y				Y				Y	Y		Y	Y	Y			Y
c. threats		Y				Y				Y	Y		Y	Y	Y			Y
2. Functions																		
a. current		Y				Y				Y	Y		Y	Y	Y			Y
b. wishes/needs						Y				Y	Y		Y					Y
c. bottlenecks										Y	Y		Y	Y				Y
3. Alternative plans																		
a. development										Y	Y			Y				Y
b. evaluation											Y			Y				Y
c. criteria for evaluation																		

Y = apparent in plan

Table 11. Research

	Estuary																		
	Huelva	Ave				Loire	Venice			Solent				Eastern Scheldt				Western Scheldt
Plan	P1	MP	P1	P2	P3	MP	P1	P2	P3	MP	P1	P2	P3	MP	P1	P2	P3	MP
1. Environmental research						Y												
a. biotic features																		
- vegetation		Y								Y	Y		Y	Y				Y
- fauna		Y								Y	Y		Y	Y				Y
b. abiotic features		Y								Y	Y		Y	Y				Y
c. landscape features										Y	Y		Y					Y
d. interrelation biotic-abiotic														Y				Y
2. Functions																		
a. shipping										Y	Y			Y				Y
b. fisheries										Y	Y		Y	Y				Y
c. agriculture										Y	Y		Y					
d. gas- and oil extraction										Y								
e. sand/gravel mining										Y	Y							Y
f. industry										Y	Y							Y
g. residential use										Y	Y		Y					
h. tourism										Y	Y		Y	Y				Y
i. land reclamation										Y	Y							
j. land embankment																		
3. Effects of functions on environment																		Y

Y = apparent in plan

Table 12. Measures

Plan	Huelva	Ave				Loire	Venice			Solent				Eastern Scheldt				Western Scheldt
	P1	MP	P1	P2	P3	MP	P1	P2	P3	MP	P1	P2	P3	MP	P1	P2	P3	MP
1. Technical measures		Y					Y											
2. Administration and organization																		
a. coordinating agencies		Y				Y	Y	Y		Y	Y			Y				Y
b. permit system						Y	Y	Y					Y					
c. penalties							Y	Y										
d. tax incentives																		
e. financial incentives		Y					Y	Y			Y							
3. Implementing measures																		
a. reduce human activities										Y	Y		Y		Y	Y		
b. continue " "										Y	Y		Y					Y
c. increase " "																		
4. Physical planning measures		Y				Y												
5. Policy																		
a. source oriented policy		Y																Y
b. effect oriented policy		Y																
6. Management																		
a. water		Y					Y	Y	Y	Y	Y		Y					
b. nature										Y	Y		Y	Y	Y			
c. landscape										Y	Y		Y					
7. Education/information		Y								Y	Y		Y	Y				Y
8. Integration																		
a. environmental policy														Y				Y
b. env. and other policies		Y								Y				Y				Y
c. implementation		Y								Y	Y		Y	Y				Y
9. Procedures																		
a. evaluation		Y												Y				Y
b. adjustment		Y												Y				Y

Y = apparent in plan

Table 13. Participants and advisors

	Estuary																	
	Huelva	Ave				Loire	Venice			Solent				Eastern Scheldt				Western Scheldt
Plan	P1	MP	P1	P2	P3	MP	P1	P2	P3	MP	P1	P2	P3	MP	P1	P2	P3	MP
1. Governmental																		
a. local/municipal		I/P				P					I/P			P	I/P			P
b. provincial/state		I/P				P				I/P	I/P		I/P	P				I/P
c. regional/interprovincial		I/P				P		I/P						P				P
d. national/federal						I/P	I/P	P	I/P				I/P	I/P				I/P
e. international																		A
2. Non-governmental																		
a. academic		P				A		P		P	P			A				A
b. private/citizen							P						P	A				A
3. Economic sectors						A												
a. shipping										A	A							
b. fishery										A	A			I				
c. agriculture																		
d. industry																		
e. tourism										I/A	I/A			I		I		

I = Initiator
P = Participant
A = Advisor

planning developments supported by inventory and research projects, and subsequent monitoring programmes.

Planning experiences should then be transferred to water systems which reveal less favourable function/conflict relationships. Although the necessity to cope with these areas and the benefits gained may be considered greater in these cases, one should make certain beforehand that the planning strategy and instruments are well prepared and implemented.

As such, the approach may at least outline one of the ways of dealing with the conflict barrier situation - starting from a somewhat less conflicting environment, such as the Eastern Scheldt water system, and working one's way towards more complex and difficult situations, as the Western Scheldt example shows.

3

REVIEW OF ESTUARINE MANAGEMENT

3.1. ASPECTS OF PLANNING

In order to gain proper insight in the range of aspects that need to be considered while managing estuarine areas, this chapter first summarizes the various stages of planning. Subsequently, attention is paid to the definition of the planning boundaries of an estuarine area.

3.1.1. Planning Process

The planning process for integrated estuarine management is linked to policy development and implementation in order to address actual and potential problems in estuarine management. This process involves the following stages (OECD, 1991, pp.74-75):

1. Detailed inventory and assessment (information and analysis) to produce the data and findings to assist policy development and implementation (thus helping to correct policy deficiencies);
2. Coordinated analysis to identify those elements in development plans which would comply with the criteria laid down and enable the modification of plans accordingly (and thus helping to correct intervention deficiencies and market failures);
3. Coordination of policy formulation, to eliminate conflicting policies being developed (thus helping to correct intervention deficiencies)
4. Establishing implementation, monitoring and evaluation mechanisms and defining responsibilities.

Systematic analysis of the natural and economic systems, and policy implementation through the use of certain instruments are an essential part of the management strategy. But neither analysis can be carried out properly, nor the various integrating instruments efficiently applied, unless the necessary institutional mechanism exists and operates successfully (OECD, 1991, p.75).

The sequence for the coastal zone management process can be presented in the following way (OECD, 1991, p.75):

1. Establishing administrative/political coordination and the creation of the institutional mechanism;
2. Generation of information (analysis/planning);
3. Reassessment of present policies;
   Reassessment of legislative requirements;
   Reassessment of legal/juridical action;

4. Preparation of alternative options and analysis of implications (environmental, social, economic) and risk/uncertainty;
5. Selection of the final plan, involving public participation;
6. Implementation;
7. Monitoring and evaluation to feedback into planning.

Conscious action in estuarine areas makes it desirable to make a number of coherent and non-conflicting pronouncements regarding the direction in which functions should ideally develop. Such pronouncements should be incorporated into a policy plan, which maps out the desired line of development. This policy plan should span the entire cycle of the planning process (Saeijs, 1982, p.233).

3.1.2. Planning Boundaries

Since it is the total area of the estuary that is taken as the basis for planning, the planning boundaries are determined in first instance by the limits of the system that is being planned (Saeijs, 1982, p.233). But how should the boundaries of such regions be defined? Factors to be considered include the boundaries of natural ecosystems, the spatial extent of economic activities, the spatial extent of nonrenewable resource deposits, and political jurisdictions. According to Bower c.s. (1982, pp.14-15) at least five different types of economic natural systems regions exist which are relevant to establishing regional boundaries for management.

One is the Waste Source - Waste Control Region, where the discharges of wastes are made and control over the discharges and over some related environmental impacts might be exercised. A second type is the Effects Region, which is a natural system of some type, a coastal water body, an ocean water mass, a fish habitat area. This type of region encompasses the spatial extent of adverse impacts resulting from wastes discharged in a source-control region. A third type is a Nonrenewable Resource Region, an area of concentrated mineral or oil and gas deposits. A fourth type is the broader Economic Region, which may be affected by management strategies affecting economic activities in the source-control region, which in turn also affect natural systems.

A fifth type is the Governmental (Political) Region, which is based on boundaries of governmental jurisdictions. These boundaries, the normal decision-making units for coastal zones, seldom coincide with ecological units; nor do they recognise the ecological units of the zone. With very few exceptions the ecological system of the coastal zone is inherently compatible with the political and administrative structures that have been developed to serve the social system. Consequently to manage coastal resources which encompass the appropriate ecosystem the various administrative units responsible for management have to be integrated. This is indeed one of the main reasons for integrated management: to achieve coordination amongst the various local and regional government units which administer parts of the ecosystem. For estuaries this integration would cover the whole area of estuarine waters and also outside the estuary to the extent that activities have mutual impacts, i.e. out from the estuary and on the estuary from the open sea (OECD, 1991, pp.34-35).

There are cases in which the area thus defined might be far too large to be treated as a coastal zone management unit, e.g. if a major river running into an estuary has a very long basin and it would be impractical to include all the various local government units. In such cases integration with those authorities will still be necessary but only to the extent of their impact on coastal waters (OECD, 1991, p.34).

Finally it needs to be mentioned that it is not necessary that mutually exclusive boundaries be defined for a specific case. Boundaries of effected regions, economic regions and political regions could overlap among management regions depending on the problem context. Nor is it necessary that economic activities and discharges of wastes and the related effects be analyzed with the same degree of detail within each type of region (Bower c.s., 1982, p.15).

3.2. ORGANIZATIONAL FRAMEWORK

The great number of organizations involved in the management of estuarine areas has led to a lack of coordination in many instances. Efforts to improve the quality of the estuarine environment therefore have not always been very successful. Many of the pressures and activities are not controlled by the planning authorities. Instead, a multiplicity of other organizations exercise separate control over pollution, fisheries, mineral extraction, navigation and so on. Often management efforts were initiatives of separate organizations and could not contribute to an integrated approach to solve the problems. There are many examples of this situation, like the case of the San Francisco Bay (USA) which Grenell (1991, p.503) described as "a patchwork of agencies, each with different mandates, functions, jurisdictional areas, and governmental positions which has, not surprisingly, resulted in duplication, inconsistencies, and often serious gaps in regulatory administration and planning."

For a better understanding of the present organizational situation around estuaries in this chapter first a summary of organizational arrangements is given. Next attention is focussed on the various parties involved in these organizations. Finally, some conditions for the set-up of an organizational framework are presented and two major administrative models are considered.

3.2.1. Present Organizational Arrangements

In Europe the responsibility for administrative procedures within the coastal and marine areas traditionally rests solely with the central government. What these governmental units are lacking, due to their purely sectoral orientated tasks, is the ability to overcome the existing fragmentation in jurisdiction and authority (Scharmann, 1990, p.539).

On the basis of a study on the performance of OECD-countries in coastal zone management Juhasz (1991, p.598) concludes that very few coastal areas created a specific integrated management framework for managing coastal zones. Partial examples are the Chesapeake Bay (USA), the Fraser Estuary (Canada), the Great Barrier Reef (Australia), the Lagoon of Venice (Italy) and the Eastern Scheldt (Netherlands).

At the other extreme Juhasz signalizes coastal areas being managed by only one or two local authorities with very little administrative authority and discretion in land zoning and other resource use decisions. An example is the Messolonghi-Aitoliko Wetland Area (Greece). According to Juhasz the in between arrangements are of two major types:

- Economic development is driven by national policies. Often there is no formal arrangement between local, regional and national authorities and environmental concerns are not an integral part of resource planning and project implementation;
- Economic development is driven by local governmental objectives and national authorities find it difficult to incorporate environmental objectives into the policy framework.

According to Juhasz a number of effects flows from these frameworks:

- Either economic development plans exclude environmental considerations and the environmental impacts are treated ex-post with consequently higher pollution control cost and greater damage, including lack of conservation measures;

- Or economic development plans are assessed without proper valuation of environmental damages and consequently the development scenario is not necessarily the optimal scenario when environmental factors are taken into account;
- In some extreme cases such developments have contributed to the breakdown of the ecosystem in coastal waters, including degradation of marine habitat and loss of marine life (e.g. certain coastal areas of the Mediterranean);
- Conflicts between different interest groups and resource sector agencies have occurred. As pressures on coastal resources have intensified the desirability of policy integration has been increasingly acknowledged in addition to the roles of property rights specification, resource pricing and conflict resolution approaches.

It is clear that insufficient organizational frameworks with a lack of coordination between parties involved and insufficient coordination of activities, have a negative impact on the conservation and restoration of estuaries. Though many of those involved in estuarine management have pointed out the necessity to come to integrated management, till now many estuaries in the world still suffer from the lack of such an integrated approach.

Therefore, the participants of the International Conference on the Environmental Management of enclosed Coastal Seas 1990 in Kobe (Japan) again stressed on the necessity to come to a comprehensive, integrated approach to coastal zone management to ensure economic development as well as environmental protection of these areas (The Seto Inland Sea Declaration, 1991, p.XVIII).

3.2.2. Organizations Involved

The administrative framework for integrated estuarine management involves a vertical element: international, national, regional and local governments; a horizontal element: the various governmental departments involved; a public and private sector element: government, private industry, and the public at large; and a geographical element: regional/local authorities in the relevant coastal zone and the economic drainage basin of the hinterland (Juhasz, 1991, p.597).

Below the role of the participants in estuarine management is examined.

Governments. Governments play a major role in coastal zone management, although it is generally recognised and accepted that coastal zone development is basically driven by market forces. The role of governments is to ensure first of all that coastal resources produce goods and services which are not produced by market forces (public goods), such as the provision of conservation of natural areas, preservation of aesthetic beauty. Secondly, governments should take care that a long term view is taken in contrast to the short-run profit maximisation criteria of the market. Thirdly, governments have to ensure that the externalities (pollution) created through the market forces are minimised and thereby also to reduce market generated inefficiencies (e.g. market failures) (OECD, 1991, p.71).

Some of the most intractable policy conflicts occur when they are between different levels of government. The results of national/local tensions can be genuine confusion as to who gives way to whom, reflecting in essence a different valuation of a resource from a local and national standpoint (OECD, 1991, p.52).

Private Users: Users of the estuary and the resources that affect estuarine quality (fishermen, boaters, merchant marine, industries, farmers, tourists and residents of the communities connected to the coast by land or streams abutting the coast) must find ways to build comprehensive coalitions to support the needed action to restore and protect the coastal environment (Scott, 1987, p.517).

Figure 6. Cumberland Basin is a macro-tidal estuary (mean tidal range 11 m) in a sparsely populated region of Canada. The construction of a tidal power station for the generating of electricity could cause major environmental changes.

Although the principal management responsibility of the private sector is, of course, management of development itself, there is an increasing appreciation of the overall management function (Smith, 1991, p.278). This is a result of a changing attitude of users of estuarine resources, who become more aware of the importance of the quality of the estuarine environment. As a result there seems to appear a general willingness among users to contribute to the improvement of the estuarine ecosystem.

Users are also being involved more and more in the process of decision-making concerning estuaries. The importance of this is imbedded in the impossibility for a project or plan to proceed without the substantive support of those directly affected. The most important among these groups are those who are economically tied to the coastal region. Their voices are loud and likely to be heard. Unless a project is demonstrably beneficial to those who live and work in the affected coastal area, the project will not proceed (Perret and Chatry, 1991, p.328).

General Public. The motivation for restoring and preserving estuaries, has come as well, from an aroused public (Scott, 1987, p.515). In 1982 Cunha (1982, p.537) observed a first tendency in estuarine management to ensure the populations the opportunities of intervening in the process of decision-making from the very beginning. Previously, decisions related with estuarine management were taken unilaterally by the governments involved. Participation of the populations in the decision, if any, took place at an advanced stage of the process, after fundamental decisions had already been taken. The change in this situation is among other things a result of the fact that estuarine problems are getting more serious and citizens are becoming more aware of these problems. Boekee et. al. (oral communication) investigating tourism along the borders of the Western Scheldt found that Dutch people

Figure 7. Birds in the halophytes plains of the Teacapan Agua Brava Lagoon.

living nearby would not enter the water to swim, while Belgian tourists did not hesitate to bathe in it. This sharply contrasting behaviour may be due to differences in handling the subject of pollution by the Dutch and Belgian authorities and press at that time.

It is now realised that government efforts at estuarine protection must be monitored and participated in by the public to ensure quality goals are achieved (Scott, 1987, p.517). In many estuarine management programmes the support of the population is recognized as an important factor to be considered in decision-making. The public has therefore often obtained a clear position within the organizational framework. Also growing attention is paid to information and education aiming at increasing the awareness of the public to estuarine problems.

The involvement of the public in coastal zone management emerges in a number of ways. Decisions to proceed further can be made by the public and private developers and different levels of government following an evaluation of public preferences. Public inquiries can be used as a forum for obtaining local or regional responses, public bodies and industry can arrange open meetings to inform the public of their plans and proposals (OECD, 1991, p.78).

Non-Governmental Organizations. Non-governmental groups provide important stimuli for government action. Their involvement may facilitate the 'politicization' of issues concerning these marine and estuarine areas. The role of such groups may be particularly important in the implementation of such policy (Kriwoken & Haward, 1991, p.157). This can be illustrated by the example of the San Francisco Bay (US) (Monroe, 1991, pp. 609-611) where environmental groups employ a variety of activities to address the estuary's problems. These activities include influencing legislation, involvement in government planning programmes, public education, input into the regulatory process, and when necessary, litigation. Decision-makers in government and the private sector are well aware of the increasing

Figure 8. Estuaries are nurseries where young fish grow until migration to the sea. Fisheries, like here in Spain, can be affected by the implementation of civil engineering projects. Also note: diving as a new opportunity.

influence the environmental groups have on public opinion and politicians. The groups are successful, as they may represent the sentiments of a large portion of the public, and especially when they contribute cooperatively to working groups, which may have widely differing interests. Yet, whenever safety is at risk or extremely large sums are at stake, lengthy procedures may outweigh necessary immediate measures. A provisional plan may then prove to be invaluable as is argued in chapter 4.

Scientific Community. The participation of scientists in estuarine management is extremely important, as their research is revealing the status of water quality and living resources in estuarine watersheds. According to Connor (1987, p.503) the role of scientists is particulary troublesome because of the triad of roles they often serve as (1) arbiters of the technical feasibility of a project, (2) a separate political constituency, and (3) potential contractors for the work needed to be done. Scientists are often asked to review the technical merits of a proposed project, although in polarized situations few scientists are perceived as unbiased. In addition, the scientists asked to evaluate a project's technical feasibility may later be contracted to perform the work.

The experience with a multi-disciplinary project in the Western Scheldt learns that a strict separation of research interests and political interests is also fundamental to the survival of a project. As soon as credibility is fading one will soon be in the awkward position of loosing the very right to exist. This is especially the case in a multi-disciplinary project which will produce information on conflicting interests. One should continuously be aware of interests, avoid any misunderstanding, and produce clear-cut research results (Scheele and Van Westen, 1990, p.500).

From a comparison between the situation in the Buzzards Bay, the Narragansett Bay and Long Island Sound, Connor (1987, pp. 501-511) concludes that three sorts of activities are key to the development of good contacts between management agencies and the academic scientific community.

Firstly, the importance of building a trusting and personal relationship between scientists and managers is emphasized. This should avoid problems in communication and minimize competition (e.g. for funds) between the agency and academic scientists. This structure of trust must be built afresh between each different agency department and academic department, as ties between one university department and agency branch are no guarantee of a good relationship between any other academic department or agency branch.

Secondly, scientists must respond actively to the managers' needs. A mechanism must exist in the academic organizations to permit and even encourage the participation of scientists in estuarine studies. The major challenge in soliciting scientific participation is enlisting scientists who are willing to focus on problems beyond their individual research interests. Therefore, the scientific community needs to take an active role in the negotiation between scientists and the agency.

Thirdly, responsive action by the managers to the scientists is needed. Connor concludes that the single most important action by the agencies which established their credibility with academic scientists was the appointment of established scientists from academic institutions to coordinate the project.

3.2.3. Conditions for the Set-Up of an Organizational Framework

Bringing together the parties involved in an organizational network should be seen as a major goal in improving the management of estuarine areas. However, in scientific literature concerning estuarine development the way in which this integration should take place has gained relatively little attention. In most cases recommendations appear to be limited to a call for "improved cooperation of the participants involved" or a "basin-wide, integrated approach to environmental management problems." A comprehensive analysis of the preconditions and possibilities to accomplish such integrated management in estuarine areas has not yet taken place very often.

Of importance here is the work performed by the OECD (1991). The OECD presented a conceptual framework for integrated coastal management based on 17 country reports, case studies and literature, most of which concerned estuarine areas. The framework suggests guidelines for formulating problem specific political and institutional responses for carrying out a number of functions broadly similar in most countries and under different circumstances. For any specific problem area these can then be broken down into clearly defined local strategies. This framework does not necessarily involve the creation of new agencies, but will involve a re-evaluation, and possibly redefinition, of the objectives, responsibilities and operation of existing governmental and quasi-governmental bodies. This is best conceptualised as the creation of a new "institutional mechanism."

The OECD-report includes in this mechanism the political/legislation, the bureaucratic/administrative and the legal/judicial sectors. An improved institutional mechanism for coastal management will involve improving the linkages among these sectors and the development of better regulatory and economic instruments. The necessary bureaucratic/administrative links would involve collaboration among representatives from national, regional and local governments of the coastal zone areas concerned, designed to develop policies for specific regions or local areas.

Although this will not necessarily entail the creation of new bodies, there will almost always be the need for a coordinating body or bodies to generate information, reassess policies, and carry out the analysis and planning for the political decision-maker. This coordinating body could consist of a management council and various sub-groups, and it should fit as far as possible the boundaries of the coastal zone area in question, to cover the main linkages between water and land resources in that coastal zone. The coordinating body might be created on an ad hoc basis to cater for the local or regional needs, although it would usually be preferable if there were a national system for coastal zone administration which then created regional and/or local bodies which fitted within that national system.

The establishment of such a coordinating mechanism should be initiated by the authorities either at the national level or at the regional level. A primary requirement, therefore, is that governments take the initiative to establish such a coordinating mechanism. This will require political will and public support, involving the politicians, consultation with industry, media interest and pressure group activity. The establishment of such coordinating bodies would involve drawing together different national departments with an interest in the coast as well as a number of regional and local authorities of an area where coastal land waters need to be managed as a unit. In the case of estuaries these areas present themselves relatively readily.

As the coastal areas involved cover both land and coastal waters those authorities responsible for the operation, exploitation, conservation and maintenance of resources in these areas should be represented. It implies that a substantial number of authorities would be involved, but they need not be participating at all levels at all times. According to the nature of the local and regional coastal problems or the nature of the proposed development projects, specific agencies might take the leading role.

For the support of integration in general and the creation of an institutional mechanism in particular, a consistent government policy at national or regional level is of primary importance. There are two reasons for this. Firstly, these higher levels of government have the financial power and specialised technical expertise and it is usually their duty as part of the overall management of the economy and environment to ensure integration. Secondly, national and regional governments are less subject to often intense local pressures to create employment and generate local economic development that can have adverse effects on a balanced coastal zone policy.

National or regional governments can create appropriate coordinating bodies by legislation, through the initiative of major government agencies responsible for the development and/or conservation of the environment, or through fiscal incentives provided on the basis of cooperative management. There are cases where local non-governmental initiatives have played an important role in the establishment of a cooperative management approach. Such initiatives could be encouraged in the future and could be supported by governmental action.

3.2.4. Institutional Arrangements

For the implementation of a management strategy in an area-based approach covering a number of linked and overlapping sectors, two major administrative models (and a series of variations on these two themes) can be considered (Saeijs, 1982, pp.208-224; Rijsberman, 1988, p.17).

In the 'Complementary Administration Model,' all authorities agree to coordinate their actions (and sub ordinate these to a jointly accepted plan) but retain their decision-making power. Coordination is organized on a voluntary basis. In practical terms, complementary administration implies the following consultation between the administrations of separate bodies at various levels, which requires carefully planned procedures on the basis of equality backed by guarantees, with clear legal consequences resulting from the consultations.

In this model all partners in decision-making can have their own executive responsibilities and organizations, but it is also possible that one of the partners of a joint executive body is responsible for implementation. In the latter case, the set up of a joint executive body, offers the advantage of the work being performed clearly and unambiguously.

The second administrative model, the 'Joint Regulation Model,' a higher administrative level is created in order to make the cooperative links more susceptible to control. This "umbrella body" can be given powers of sanction in order to make the venture less of an open-ended commitment, as well as the power to pass regulations and its own source of income. The authority regarding a specific area is thus transferred from all other government agencies to a new agency.

In many cases the Complementary Administration Model is easier to implement, because coordination is done by bodies which already exist. The Joint Regulation Model presents practical problems associated with the creation of a new government authority and should be regarded as a long term solution.

3.3. POLICY ANALYSIS

3.3.1. Policy Cycle

Policy analysis is an important link in the process of formulating policy on estuaries. It is part of a similar policy cycle which is also applied in a wide range of social issues. Winsemius (1986) identifies four stages:

1. Problem identification
2. Policy formulation
3. Policy implementation
4. Management

Problem Identification. This is the stage where concern is voiced that a problem exists. For some period, however, a degree of uncertainty still exists as to its nature, extent, causes and effects. This stage of the policy cycle comes to an end when it is generally agreed that a problem is manifest. The problem is then identified as a policy issue. This is usually the case once we have a better understanding of the sources, risks, effects and possible solutions. Disasters tend to accelerate the process considerably.

The organisation responsible for seeking a solution has also usually been identified by the problem identification stage. In the case of estuarine management, this will usually be the local water authority.

Formulation of Policy. Policy formulation takes place at two stages; the initial groundwork of policy preparation, and then the choice of an alternative.

In policy preparation, the accent is on careful formulation of the problem and finding solutions. Policy is prepared in four steps:

- policy analysis, which is aimed at:
  - formulating the problem in terms of a desirable situation to be striven for;
  - defining the extent of the problem by comparing the existing situation with the desirable situation and
  - finding solutions to achieve the ideal situation or a near approximation to it.
- procedures: These are the rules of the game to be observed during the policy formulation process. They relate to powers, the time available for consultations and recommendations, and the involvement of community-based groups. Procedures can strongly influence the direction in which policy formulation develops. They may be of a statutory nature, as in countries which require Environmental

Figure 9. Problem identification is not time independent. In 1970 the Haringvliet estuary was closed off by discharge sluices and changed in a fresh water lake to safeguard safety and to improve water management. In 1989, the loss of the ecologically important brackish water zones was recognised as a serious problem in the Third National Policy Document on Water Management.

Impact Statements. In other cases, new procedures might be designed for each policy problem.

- consultations and recommendations. Consultations are held with community-based groups and any statutory advisory bodies before decisions are taken. If any, public information campaigns are usually also conducted at this stage (see section 3.5.5).

When policy preparation has been completed, those responsible for drawing up policy have to make a choice. The policy analyst has collected the relevant information and rank-ordered to help the policy-maker reach a deliberate decision. It is up to the policy-maker to examine the alternatives in the light of objectives, general policy and the public interest, and to make a choice. The policy analyst will frequently indicate the procedure which should be followed. It is vital for all concerned that sound reasoning should be put forward for choosing a particular alternative.

Policy Implementation. The chosen solution, which usually comprises a package of measures, is put into practice at this stage. The focus here is usually on the groups responsible for implementing the measures. Public interest often diminishes substantially at this stage.

Management. Management can be defined as those activities which are geared to ensuring that all estuaries meet the requirements of policy — either by continuing to do so, or by meeting them at some future point in time (Rijkswaterstaat, 1993). Day-to-day

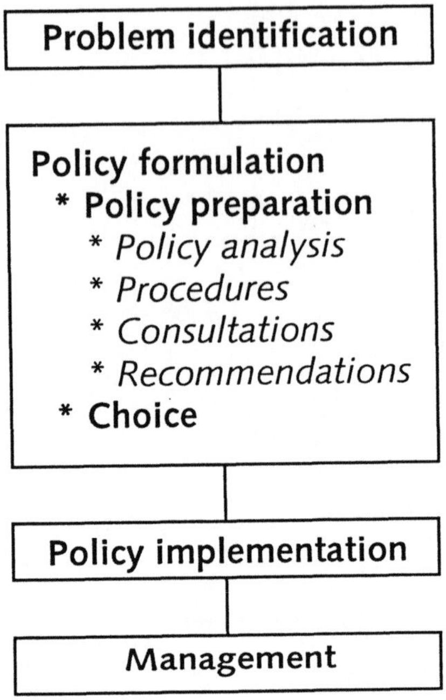

Scheme 1. The position of policy analysis in the policy cycle.

management of the estuary may go relatively unnoticed, but is nevertheless essential if policy is to be put into practice. This involves numerous activities of varying nature. Property management, for instance, involves operating and maintaining the physical infrastructure, such as locks, weirs, banks and canal beds. Other aspects of management relate to the enforcement of legislation and regulations, including the issue of licences.

Effective management is, of course, first and foremost the responsibility of the authorities concerned. Where special interests are involved, the interested parties generally share responsibility for both organisation and funding. Management might then include the demarcation of areas reserved for nature conservation by means of markings for shipping and the creation of joint bodies to share responsibility and costs.

Evaluation is not included as a separate stage in Winsemius's policy cycle, as it is regarded to be an integral part of the management stage. In many cases, however, evaluation is seen as a separate stage with its own procedures (Van Westen and Colijn, 1993).

The rest of this section will deal mainly with policy analysis. Its position within the policy cycle is shown in Scheme 1 (Rijkswaterstaat, 1988).

3.3.2. Policy Analysis

There are many definitions of the concept of the Policy analysis. A short and clearcut definition frequently used states that Policy analysis facilitates choice. Its most important objective is thus to collect, arrange and present information, and not — as the above shows — actually making choices. Policy analysis may be performed to prepare either a policy or a project. For this reason it is often called a policy study or project study. Both kinds of study can adopt a similar approach, referred to here as the policy analysis approach. In the policy analysis approach, alternative solutions to a policy problem are systematically developed

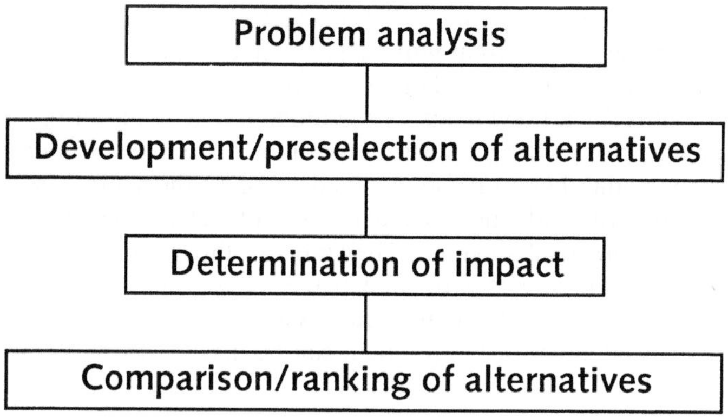

Scheme 2. Steps in policy analysis.

and evaluated. Several steps can be distinguished, as shown in Scheme 2. These will be discussed in more detail later.

Such a systematic approach has a number of important advantages. The policy study or project study can be completed relatively quickly, is manageable and testable, and clear to all involved. Also, uncertainties — present and future — can usually be assessed more accurately, although it should be emphasised that they cannot be eliminated by a systematic approach.

When should a policy analysis approach be pursued? The area of application is very wide and ranges from nebulous political and social problems to extremely concrete technical issues and from national policy problems to internal staff matters.

A policy analysis approach has advantages, especially if:

- public interests at stake
- these interests conflict
- quantitatively dissimilar interests are involved
- a large number of quantities have to be compared

in short: if the choice is complex

3.3.3. Implementation of Policy Analysis

General. Scheme 2 shows a number of steps in policy analysis. In practice, however, it is not that simple. In the first place, the different stages can be subdivided into other component activities. Then a suitable method or model has to be chosen for each activity. Finally, as a rule the process does not run smoothly from beginning to end, but frequently doubles back to an earlier stage — in other words, the process is iterative. For example, there might turn out to be more problems than were envisaged when the alternatives were being assessed, or more than one solution might be possible.

A more detailed description of the steps of policy analysis is as follows.

Problem Analysis. At this stage, the policy problem is stated in as concrete a way as possible. This involves defining the difference between the present (or forecast) situation and some ideal situation.

Determining the extent of the problem usually means carrying out a study on its underlying causes and assessing what will happen if nothing is done (the null alternative), including the likely consequences for those affected.

The study aims for a knowledge of the processes at work on the one hand, and a thorough understanding of public interests on the other. To arrive at a practical statement of the problem, it is essential that all those involved agree on the standards and objectives to be used at this stage. After all, these standards and objectives will be used as objective quantities to represent the ideal situation. They are also the means of evaluating the extent to which predicted developments will take place. It is important, therefore, for these standards and objectives to be defined at an early stage.

Two kinds of statement of the problem can be derived from the problem analysis, that is, to the question:

1. is more work needed or does a policy need to be developed?
2. which is the best alternative?

It goes without saying that if the answer to the first question is yes, then the second question arises automatically. It might then be necessary to carry out the problem analysis in more detail.

Developing Alternatives. The development of alternatives involves developing solution areas — sometimes a great many — based on an understanding of the policy problem. In analogy with the problem analysis, solution areas are first defined, and then a limited number of alternatives amenable to analysis are chosen. An alternative is defined as a set of measures which are sufficient to solve the problem. Evidently, an alternative may consist of just a single measure. Variants can produce different possibilities within a single alternative.

The development of alternatives is not simply a technical matter — a hard and fast analytical method is not recommended. Next to considering the problem itself, important aspects to take into account, include:

- solutions already proposed
- experience in similar situations
- causes of the problem
- any constraints
- the urgency with which the problem must be solved.

A number of useful methods can be suggested to arrive at a reasonable set of alternatives in a more or less systematic manner:

- 0 alternative
- 0+ alternative
- analogous solutions
- tackling the cause
- brainstorming

Preselection. It is highly probable that we will have to deal with a large, and possibly excessive, number of alternatives at the preselection stage. It may be impractical and expensive to elaborate, to present and ultimately to evaluate all the possible alternatives. For this reason, preselection is carried out where necessary. A large number of alternatives is reduced to a manageable number of preferred solutions. Preselection is usually done without setting rigid criteria. Reasons for inclusion might be:

- cost
- feasibility (technical, political or social)

- attractiveness (the problem solving capacity, or the degree to which the envisaged effect will be achieved)
- dominance (some solutions are clearly better than others).

Evaluation Criteria. Before effects can be predicted, a choice of evaluation criteria has to be made. In principle, evaluation criteria are selected in such a way that the effects are able to influence the ranking of alternatives. This means that evaluation criteria must have sufficient weight, must have the capacity to discriminate, and must be independent:

- sufficient weight implies that the predicted effects will be of a certain size;
- the capacity to discriminate implies an ability to generate differences between alternatives;
- independence implies that there is little or no overlap between evaluation criteria.

Determination of Effect. Several different kinds of effect can be distinguished:

- direct and indirect effects. Direct effects arise as an immediate consequence of the implementation of an alternative; indirect effects arise from direct effects.
- temporary and permanent effects.
- reversible and irreversible effects. For reversible effects, measures can be taken to reduce their seriousness or extent.
- internal and external effects. Internal effects have an impact on the organisation taking the decision; external effects are unintentional effects experienced by third parties.
- intended and unintended effects.

Intended effects are similar to the objectives already identified in the problem analysis. For coastal defences, for instance, these might be a particular dyke height or an ideal width of beach. These effects are the conditions which the alternative must meet. Unintended effects are those which occur anyway, but do not contribute to achieving the objectives of the project. These can be either beneficial or adverse. The fact that such effects occur, and that they usually have impact on conflicting interests, underlines the need for a policy analysis.

Determining these effects is not always straightforward; sometimes knowledge is lacking, and it is costly and time-consuming to carry out the necessary research. A point to note here is that these effects frequently have impact on matters beyond the jurisdiction of the principal, who is nevertheless responsible for choosing the aspects to be evaluated and the scope of the effects. It goes without saying, therefore, that full consultations should be held with all those involved.

Many methods are available for determining effects. Examples include numerical models, statistical processing, dose-effect relationships, process descriptions, expert judgements and expert systems. These methods are not dealt with here, since they are adequately covered in literature.

The process of forecasting effects cannot be separated from the concept of uncertainty. The models used are fundamentally only approximations of reality, which itself is liable to change in unpredictable ways. Ways of reducing uncertainty include:

- collecting more data and doing more research
- flexibility in design
- performing sensitivity analyses to yield an insight into the reliability of estimates. In any case, uncertainties should always be included and explained in the final report.

Comparison and Ranking of Alternatives. Data collected has to be structured and presented in an orderly way. This can be done using an effects summary of both quantitative and qualitative information. In this sense, "qualitative" means that the effects are evaluated in the light of a given reference situation, or that the effects of different alternatives have been compared.

Whether the null alternative is included in the summary depends mainly on the statement of the problem. If the issue is whether or not to implement a project, the null alternative is appropriate. If the aim is to determine the best alternative, the null alternative can be presented as a reference, but this is not obligatory.

Evaluation Methods. The effects summary can conclude the project study. However, depending on the problem to be solved and the nature of the interests to be considered, further processing of the material may be required. In principle, a whole range of evaluation methods is available. But most of these methods are not so universally applicable as was hoped for, and in some situations a separate study is called for to decide the best method. Some methods are computerised and capable of handling extremely complex problems.

Some of the methods which can be used to solve conflicting problems are discussed in more detail below. We can distinguish three main groups of methods:

1) monetary methods

These include cost-benefit analysis. This will not be discussed further here, except to say that a precondition of its use is that all effects must be capable of expression in money terms.

2) summary tables

The best known of these is the score-card method described below.

3) weighted summation

This is used in multi-criteria analysis.

The score-card method entails using a code — usually a colour — to represent the ranking of alternatives against each evaluation criterion. This is always done in such a way that the original scores remain visible. The result is a visual impression of the effects summary. A feature of this method is that the policy-maker arrives at a choice himself by assigning weights to various evaluation aspects. An immediate decision is only possible if one of the alternatives is clearly dominant.

A method that is constantly being debated is multi-criteria analysis. A feature of this method is that policy-makers — and certainly not the policy analyst — give their preferences by assigning weights to particular evaluation aspects. A ranking of alternatives can then be given using a variety of calculation techniques, with or without a computer model. By feeding in different sets of weights, a sensitivity analysis can be performed.

The fact that weights have to be assigned is often cited as an argument against the method. This ignores the fact that weights are also implicitly assigned in other methods used to select alternatives. In these cases, they just happen to be more difficult to control, and are sometimes even inconsistent with previous policy.

Presentation. In most cases, a study will conclude with a final report, project memorandum or the like. Policy-makers then extract the information they require from the report and its appendices. Depending on the policy area or the project under consideration, a great deal more can be done to provide the policy-maker with information. Examples might include slide shows, field trips or visits to similar projects.

3.4. OBJECTIVES OF ESTUARINE MANAGEMENT

In an estuary a variety of uses or functions may take place. In any management plan it is therefore necessary that specific objectives for the estuary are determined (Wood, 1982, p.492). This chapter focusses on the problem of defining correct and clear management objectives for estuarine areas.

3.4.1. Policy Aims

Successful programme implementation depends on the clarity of goals. Vaguely worded goals are serious obstacles to programme implementation (Hildebrand, 1989, p.23). Since it is clearly impossible to please all interests all of the time, the goal of management must be to optimize the use of the estuary, in other words to please as many of the interests as much of the time as possible (Wilson, 1988, p.144).

One could question if the eventual goal of the management plan should be a first class, totally unpolluted estuary. According to Wilson (1988, p.145) given the present state of many estuaries such an objective would be totally out of the question at least in the short term, even if the will and the finances were there. Therefore secondary quality objectives should be adopted, tacitly accepting some degree of contamination. While this may be unacceptable to some, it is obviously a more pragmatic approach. Saeijs (1982, p.57) stated in this context that a pragmatic approach via functional objectives, criteria, standards and planning should be combined with a theoretical approach to the system by the formulation of objectives, criteria and standards for an ideal estuary of a given type or for the design of a model estuary.

Figure 10. Policy formulation should include all functions present.

When objectives, criteria and standards are considered and formulated for estuaries, the factors taken into account are almost always limited to primary social needs, or to functions with immediate social relevance such as commercial and recreational fishing, aquaculture, drinking-water supply, recreational swimming facilities. This attitude is understandable and even recommendable in terms of motivation, but however legitimate these objectives and demands may be, much more is involved, i.e. the integrity, the diversity and the survival of the environmental type of the estuary whether or not modified or manageable with advanced technical tools (Saeijs, 1982, pp.55-66).

To protect resources environmental quality objectives must be set for each estuary system. According to DeMoss (1987, pp.23-24) four different options emerge for current uses of estuaries:

1. Status quo; this option continues the managerial and administrative programmes within present resources. However, this option ignores increasing conflicts associated with population growth. Maintaining the status quo will most likely lead to degradation of environmental quality in the estuary due to increased pollution loads from growing populations, industrial development, water use demands, and habitat modification.
2. The second option is to maintain and protect resources the way they exist right now in the estuary. Maintenance of present environmental quality will require action today to mitigate the impacts of continued growth and development in the watershed.
   It will require better integration of existing resources, but will also require new initiatives and changes in current practices.
3. The third option would actually maintain current environmental quality for parts of the estuary and restore some targeted areas to a previous desired condition. This option will require even more intensity of new initiatives.
4. The fourth option is a combination of maintenance of current resources, restoration of parts of the system, and restoration and/or maintenance of some parts of the system in a pristine condition.

These different options could serve to consider tentatively several scenarios to study how improvement of the environmental quality could alleviate the constraints of the various functions in an estuary. An example of such a scenario approach is found in the Western Scheldt policy where three water quality scenarios were considered ranging from unchanged policy regarding effluents to complete elimination of pollution (Scheele, 1991, pp.203-204).

3.4.2. Environmental Standards

In order to define broad policy aims more precisely, specific information is necessary. Such information may be expressed by way of Environmental Quality Objectives (EQO): the uses to which the receiving environment is legitimately put. At practical level, the objectives are achieved by the application of Environmental Quality Standards (EQS): the maximum permissable concentrations of potential pollutants in the receiving environment, and discharge consent conditions on effluents (MacKay, 1982, p.163). Thus, the EQO are stated aims for an estuary, whereas the EQS are the conditions used to bring about those EQO.

A typical set of Environmental Quality Objectives would be that the estuary should allow (McLusky, 1989, p.178):

1. The protection of all existing defined uses of the estuary system.

2. The ability to support the biota on and in the bottom necessary for sustaining sea fisheries.
3. The ability to allow the passage of migratory fish at all stages of the tide
4. Low chemical and microbial contamination of the biota, which should not affect its consumption by man or other organisms

The above list has been chosen mainly on biological grounds, but there could be an EQO for each defined usage of the estuary. For example, water sports might require a different list.

The EQO and EQS do not have to be the same all over the estuary. The example of the Tejo estuary (Portugal) shows a division into several homogeneous zones, each with specific uses. As a consequence, different EQO are achieved, according to the appropriate water quality criteria and standards (Cordosa da Silva and Castanheiro, 1985, p.3).

The adoption of EQO has been criticized by several agencies and governments as being unjust. The opponents of EQO say that they unfairly penalise a town or an industry situated inland or near the head of an estuary, where there is little scope for dilution of an effluent, against a town or industry situated on the coast, or near the mouth of an estuary where there is abundant scope for dilution available. Thus more lenient EQS might be applied to the discharge from an industry at the estuary mouth.

To overcome this criticism, a set of numerical standards can be adopted, to which all industries or towns must conform. These standards, known as Uniform Emission Standards (EUS) are applied directly to all effluents irrespective of location, so that they are clearly different from EQO and EQS which relate to the quality of the received waters. The EUS approach is to establish a national or international standard for discharge of a pollutant, and then to apply those standards to every discharge, irrespective of the location of the discharge. The emphasis is clearly on the uniformity of discharge from different sources or locations (McLusky, 1989, pp.178-180).

When the policies of management in estuaries are implemented as planning objectives, the concepts of EQO, EQS and EUS are perhaps not as different as they first appear. In practice they usually work together, in that the estuarine manager needs to specify the permitted standards in any effluent entering the estuary. Whether the standards set are uniform in all countries, or parts of that country, is essentially a political decision. The role of the estuarine scientist is to advise the politicians as to what standards are required, either in the effluent, or in the received waters, so that the estuary is fit for its uses (McLusky, 1989, p.185).

3.5. INFORMATION FOR ESTUARINE MANAGEMENT

Knowledge is essential for estuarine management. There is an expanding need for the scientific study of estuarine systems to provide the basis for the advice required by the administrative authorities which have the responsibility of determining whether proposals, which could affect estuaries, should be rejected or approved, and in the latter case what modifications may be necessary to minimize any possible adverse effect (Allan, 1983, p.2).

Because the ultimate success of any restoration effort will depend upon the extent of the data and the quality of research facilities focussed on the estuary, adequate management should include attention for research programmes (Scott, 1987, p.517). Such research programmes should be designed to increase the scientific knowledge about the estuary and to apply this information to the management of the area. They must not only cover the technical and natural sciences, but must also be economic and social oriented (Saeijs, 1982, p.189).

In this chapter we will first have a closer look at the sorts of information needed for estuarine management. After that, the role of information when it comes to monitoring and evaluation is analyzed and attention is paid to the use of models as a tool for providing information for decision-makers.

3.5.1. Information Requirements

According to DeMoss (1987, p.27) two levels of research should be distinguished. The first level of research requires some answers now, or in the immediate future, to pressing management questions. To answer these questions one should build upon and use research that has been done over the past 15-20 years and properly assess and synthesize it. The second level is the generation of new, basic information on estuaries, which will assist resource managers in the future.

Information should be generated as well on aspects of the natural system as on aspects of the economic system (OECD, 1991, pp.87-92).

Information on the Natural System. The quantity of the stock of resources is as important as the quality of the natural system. Consequently criteria need to be laid down for the rate of consumption and restoration of resources. It will also be necessary to establish the present rate of resource use based on the current economic pattern and likely future developments. This information will indicate the quantitative resource requirements of future economic scenarios.

Also information is needed on the amount of pollution. In order to assess waste generated in the coastal region it is necessary to add waste produced in other regions and discharged into the coastal zone by rivers and by air borne pollutants. The information and measurements on waste flows calculated on the basis of existing economic activities, and measurements of ambient quality, provides the basic information on the natural system of the coastal zone. Similar calculations on waste flows can also be made for future development scenarios.

Conservation requirements should also be part of the environmental information inventory and be taken into account in the planning process. In addition to certain conservation and preservation requirements specified in the national legislation, certain requirements need to be laid down on a regional basis.

The analysis of the natural system should include:

1. physical coastal processes and the implications of interventions /alterations on those processes.
2. demands on coastal resources: from existing activities, from proposed developments, the cumulative impact of developments.
3. natural resource protection requirements: conservation, preservation.
4. externalities from economic activities: pollution and damage from existing activities and from proposed developments.

Information on the Economic System. Both from the economic as well as from the environmental point of view, detailed information is needed on basic economic activities occurring and anticipated in the coastal region. For long-term management, the first step is to prepare an economic inventory indicating the main industries of the region, including agriculture, tourism, etc. and the major inputs into industries.

Physical and economic data concerning the coast, including coastal waters, are also needed. This includes information on the economic uses of land resources, the economic uses of coastal waters, including those resulting from reclamation activities, and the economic uses of marine resource stocks.

Basic economic information is needed on proposed and future activities for two reasons. Firstly, it is needed to evaluate the amount of waste generated by the economy. Secondly, it is needed to assess the compatibility of present activities and possible economic development in the region. A proposal for further economic development needs to be judged on the economic viability of the proposal, the compatibility with other activities in the region and overall environmental impacts.

The demand on resources should be evaluated in the light of alternative developments. For this reason it is useful to assess more than one project type and economic scenario. Also alternative (low resource intensity) technical scenarios to produce the desired output should be considered in the examination of a specific development plan.

The analysis of the economic system should include:

1. existing economic activities (on a regional economic basis).
2. demands on resources of the coastal zone.
3. proposed developments for the area (single projects, development scenarios).
4. an economic evaluation of proposed developments.

3.5.2. Monitoring

An estuarine programme should always be complemented with a system of monitoring and evaluation acting as a feedback mechanism, to allow adjustment in the course of action (Coccossis, 1985, p.24; DeMoss, 1987, p.26). Such a system usually requires a baseline record on key features, issues and indicators, a method to assess the effects of action and a procedure to evaluate the significance of consequences on the environment, the economy and society. All new projects, plans and programmes should be screened through the evaluation scheme to assess their impacts and contribution to achieving stated goals (Coccossis, 1985, p.24).

The kind of monitoring programme required has usually to be devised as to fit the individual circumstance. Economics may be an important factor in the final decision (Wilson, 1988, p.162). Monitoring is a costly activity, particularly as it can demand resources each year on a continuing basis. The monitoring programme should therefore be reduced to the minimum required to ensure that the quality standard is being achieved (Wood, 1982, p.497). Monitoring needs can be so expensive and extensive, that they may have to go beyond governmental programmes and include citizen volunteers (Grigg, 1990, p.5).

A partial list of characterization parameters, according to Grigg (1990, p.5) would include: physical parameters (land use, hydrology, shoreline development, erosion rates and storm events); chemical parameters (nutrients, dissolved oxygen, phosphate, total nitrogen, inorganic nitrogen, nitrate, ammonium, organic nitrogen, toxic metals, pesticides and organics); and biological parameters (landings, catch per unit, nursery areas juvenile index, spawning areas, and plant data). According to the OECD (1991, pp.65-66) also social parameters could be considered (density of population, protection of sites of cultural and archaeological value, and ratio of development to undeveloped land).

Two examples of useful tools for monitoring and evaluation are an information system and an environmental impact assessment procedure (EIA) as presented below (Coccossis, 1985, p.24):

1. A basic requirement for any management and planning activity is adequate knowledge of the system towards which intended actions are directed. A lot of attention has been channelled toward that end. A representative case is the Information System on the State of the Environmental and Natural Resources

proposed by the EU. It is an attempt to provide a systematic framework for the collection and evaluation of data on the equality of the environment which applies well to coastal regions.

2. One of the most versatile tools developed to assess the consequences of proposed action on the natural environment, is Environmental Impact Assessment (EIA). The purpose of EIA is to identify and assess the direct and indirect effects of a project on human beings, the fauna and flora, soil, water, air, climate, and landscape, on their interrelationships, on material wealth and cultural heritage.

It is argued that EIA can help shape policies towards more desirable directions integrating environmental protection and socio-economic development.

The state of art of monitoring will continue to evolve due to the complexity of the processes to be monitored and studied. It will be important that monitoring will be designed to complement the use of mathematical models in the future (Grigg, 1990, p.5).

3.5.3. Modelling

Estuarine modelling has been shown to be an important tool in estuarine management. Although the accuracy of the quantitative predictions is still in discussion, they are very helpful in analyzing the complex ecosystem with a large number of interacting processes. Because of the highly dynamic character of an estuary and its complex structure it would be impossible to understand the main characteristics of the system and its reactions on various forcing functions. In addition, the development of such models has proved to be important in stimulating the interdisciplinary approach to estuarine research (Leussen and Dronkers, 1990, p.19; Sundermann c.s., 1990, pp.25-26).

The aim of system analysis models is not in the first place to stimulate the physical, chemical or biological behaviour of estuaries, but to support management decisions directly. Estuary simulation models are coupled to social and economic models, in order to predict the impact of measures not only on the ecosystem itself, but also on other aspects relevant to society. Often such models have a structure similar to those used in economics: input/output analysis, optimization techniques, cost/benefit analysis, etc. Mostly these are simple models, in which much attention is given to clear presentations, using modern computer graphics techniques. They meet the needs to come to a well-considered decision between alternatives and to elucidate the consequences of them (Leussen and Dronkers, 1990, p.14).

Although, according to Grigg (1990, p.6), the actual applications of models for estuarine management are limited, an increasing trend in applying mathematical models in estuarine water quality management is manifest. This trend is parallel to the rapid developments in computing facilities during the last decades (Leussen and Dronkers, 1990, p.14).

Because of the complexity of nature not all details can be quantified in a model. However, for management decisions not all these details have to be known. Therefore, each problem needs its own specific model. In relation to the trend to increase the complexity of the models, it should be emphasized that the model to be applied should be more complicated than is required to give an accuracy that is acceptable to the decision-maker. To understand the characteristics of an estuarine system holistic models are developed, incorporating all the physical, chemical and biological processes which are thought to be of importance in the given system. Knowledge of the complex interactions of these processes is fundamental to our understanding of estuarine behaviour and requires interdisciplinary studies (Leussen and Dronkers, 1990, p.14).

Models help to understand the complex relations that we cannot measure and provide information for decision-making. The first objective, to improve understanding, is rather forgiving; if the model is not completely valid, it can be excused for because it is a research tool. The second reason, decision-making, is not as forgiving, however, because decisions

Figure 11. Healthy estuaries present an enormous abundance in species, together forming a complex network.

have consequences, such as imposing costs on business, cities and farmers. With this observation it is not surprising that there are many estuary research models, but not so many that have been used for decision-making (Grigg, 1990, p.6).

3.5.4. Lack of Information

It is important to emphasise that management decisions should be seen as flexible to a certain extent, such that future operations or conditions could be reviewed or changed in the light of new information or monitored results. According to Jensen (1988, p.124) the expectation by sections of the community is unreasonable that scientists should be able to prove how an ecosystem works and to demonstrate cause/effect relationships when pollution and negative impacts occur.

Given the lack of baseline data in so many cases, it is important to change this expectation to allow scientists to present indicative information and recommendations based on trends and extrapolation, rather than spending many years to obtain more exact data, possibly too late. Scientists themselves should also be prepared to give opinions on possible outcomes without rigorous data when the need is urgent.

In general the amount of exact information will differ for the various stages of research. According to Scheele and Van Westen (1990, pp.500-501) the research level of water systems might be specified as the scientific answer to three stages of development:

1. What is actually wrong;
2. What are the effects of what is wrong;
3. What should be done about it.

Figure 12. The impacts of civil engineering projects in estuaries can be far reaching. The loss of intertidal flats in western Europe on which wintering and migrating birds feed, has a detrimental effect on the ecology of the northern tundras as far away as Siberia and Canada.

The methodology of the first stage is reasonably developed, additional research tending to be more and more marginal. The second stage is far less developed. Though it is understood that there is some relationship between ecological disasters and contamination, the causal relationships are only very partially understood: (eco)toxicology is still in its very infancy. This also implies that the third stage is a difficult one to manage.

Scheele and Van Westen conclude that it is impossible to be completely sure about the effects of certain changes or measures to be taken. Therefore politics, more than sciences, is dealing with the second and third stage of development, assuring that action is undertaken without awaiting full certainty. Since it is almost always a question of using available resources and methods in a field of knowledge which is still developing, scope must also be left for experience and common sense to come into play; in day-to-day practice one must not attach undue importance to an overly strict application of the rules of science nor must expectations of the results that science can produce be pitched too high (Saeijs, 1982, p.198).

3.6. MANAGEMENT MEASURES AND INSTRUMENTS

Policies and plans require instruments for implementation. These may be quality targets, investment programmes, licences, regulations, zoning, performance standards, etc. The accurate targeting of the appropriate instrument to solve particular problems is important for success. Such accuracy appears to have been rare in coastal management in the past, probably because of the lack of specific mechanisms for integrated management and inadequate understanding of coastal processes and impacts (OECD, 1991, p.56). Both the instruments and the authorities involved can complement each other or be mutually conflicting. This requires a sound horizontal and vertical co-ordination of the policies of the public authorities, in the field of physical planning and environmental policy, and the will to cooperate (Saeijs, 1982, p.198).

The range of instruments mentioned in literature on coastal and estuarine management can be classified under the headings of regulatory/legislative instruments, land-use planning and zoning, economic instruments, technical instruments, and communication/education. It is essential that these instruments can be operated in combination. Each actual situation must be reviewed to ascertain which instruments or combination of instruments should be applied and by whom, when and how (Saeijs, 1982, p.198).

3.6.1. Regulatory/Legislative Instruments

One of the most important aspects of government activity as far as preventive measures are concerned is in the area of regulation for setting standards for the polluters of the estuarine environment (Ojo, 1990, 114). The principles of the water resource management policy and regulations concerning the different specific aspects can be defined by the establishment of basic laws which condition the use of water resources (Cunha, 1982, p.535).

Advantages attributed to the use of direct regulations are the grip authorities have on the behaviour of actors. If there is considerable experience by authorities in their operation and the rules are clear to all parties, it provides a sense of certainty. A disadvantage is, however, that direct regulations are increasingly felt to be static, inflexible and suboptimal in terms of environmental and economic efficiency (OECD, 1991, p.98).

While regulations can prevent or limit certain actions that would injure the estuary resources, it cannot, however, take positive action to restore degraded resources or, resolve use conflicts; nor can it prevent loss of resource lands through private sales or on-site actions outside regulatory jurisdiction (Grenell, 1990, p.504).

Bargaining and negotiation processes that supplement direct regulation may offer a new direction.

The advantage of this approach is that it produces a higher likelihood of compliance than a situation of regulation without negotiation. Secondly, it introduces flexibility to adapt to new or changed circumstances (economic, technological, social) (OECD, 1991, pp.98-99).

Jensen (1988, p.126) stresses that regulation should be easily enforced and understood, avoiding complications of exemptions and conditions which confuse the public and provide loopholes to avoid prosecution. A simple prohibition is much more effective than a set of special conditions for different seasons etc.

3.6.2. Land-Use Planning and Zoning

Land-use planning and zoning makes part of regulation but is of extra importance to estuarine management because it is found to be one of the most effective instruments for separating conflicting activities (e.g. tourism and industry), maintaining a certain population density and ensuring conservation of landscape and green areas (Juhasz, 1991, p.599; Sanbongi, 1991, p.565).

This instrument is widely used at the regional level, where land is identified for dominant users (agriculture, forestry, urban, industrial), and at the local level, where more detailed land-use planning is implemented (Juhasz, 1991, p.599).

According to Juhasz (1991, p.599) the system is effective in the short run but subject to gradual erosion in the long run by a number of factors:

- The value of zoning is eroded at the edges of the zoned area as land values there increase manyfold due to zoning itself;
- Over time the high value of land leads to changes in number of building permits, and increased population density;
- Land is sometimes zoned or rezoned without public participation in decision-making and without consideration of the longer term environmental consequences;
- In some areas local authority planning and enforcement is too weak and results in unplanned development around city areas.

Some of the issues associated with misallocation of land can be corrected through more efficient land allocation by reducing governmental subsidies to particular economic activities (e.g. agriculture) which distort land prices. Also the development of side taxes can be implemented to capture windfall profits from zoning and using the tax to buy land for conservation (Juhasz, 1991, p.600).

3.6.3. Economic Instruments

Among the economic measures for the management of estuarine systems we may mention the definition on an adequate system of rates to be paid for the use of water, or the establishment of a system of charges paid for the discharge of pollution effluents, conceived in such terms that they act as economic incentives to favour better management of estuarine water resources. Other economic measures are grants, credits and exemption of taxes (Cunha, 1982, p.534).

To cover the cost of supply, depletion and environmental cost of provision, use and disposal of natural resources, a system of pricing can be established. At present some resources are either badly underpriced or not priced at all leading to their overuse and overpollution. It is acknowledged that at this stage many environmental goods and services of the coastal zone cannot be reflected in the pricing system (e.g. preservation of rare species). Governments have a responsibility to protect and maintain such goods and services for the public interest (Juhasz, 1990, p.600).

Economic instruments such as charges, subsidies, transferable permits, etc. incorporate environmental measures into economic activities (production, consumption and savings) and provide economic choices for the polluter/consumer (Juhasz, 1991, p.699). The application of charge systems will encourage users to treat until certain limits of pollution are attained and to pay for the residual pollution discharged.

The amounts received through charges paid will enable the agency responsible for the management of the estuary to pay compensations to the users who suffer greatest damage due to water pollution, to finance systems of treatment installed by some water users and to pay for collective works intended for the control of residual pollution of estuaries (Cunha, 1982, 535).

Economic instruments contain a number of advantages over regulations (OECD, 1991, p.100):

- They can be more cost effective by allowing the polluting activity to determine the most appropriate ways of meeting a given standard, or by equating the marginal cost of treatment to the level of emissions charges across the whole range of activities;
- They offer an ongoing incentive to reduce pollution below the levels determined by the regulation. They also encourage the development of new pollution control technology and non-polluting products;

- They increase flexibility. For authorities it is often easier to modify and adjust a charge than to change legislation or regulation; for polluters the freedom to choose within an overall financial constraint is preserved.

3.6.4. Technical Instruments

Technical measures vary considerably and have been used in the past with more or less success. Some examples of these technical measures are: flow regulation by means of dams constructed upstream of the estuary, river and estuary confinement by levees, dredging, control of river basin erosion or of littoral drift with the view of conditioning silting of the estuaries, and control of pollution.

It is considered that the main issue at present is not to develop new technologies but to use the available techniques better for the benefit of planners, managers and decision-makers. In this respect economic and institutional measures are of paramount importance (Cunha, 1982, p.534).

3.6.5. Communication and Education

An education programme is intended to increase resource protection in two ways: to create a better informed public by traditional environmental education programmes; to provide educational opportunities for professionals involved in education, regulation and policy development of the estuary (Thoemke, 1987, p.771).

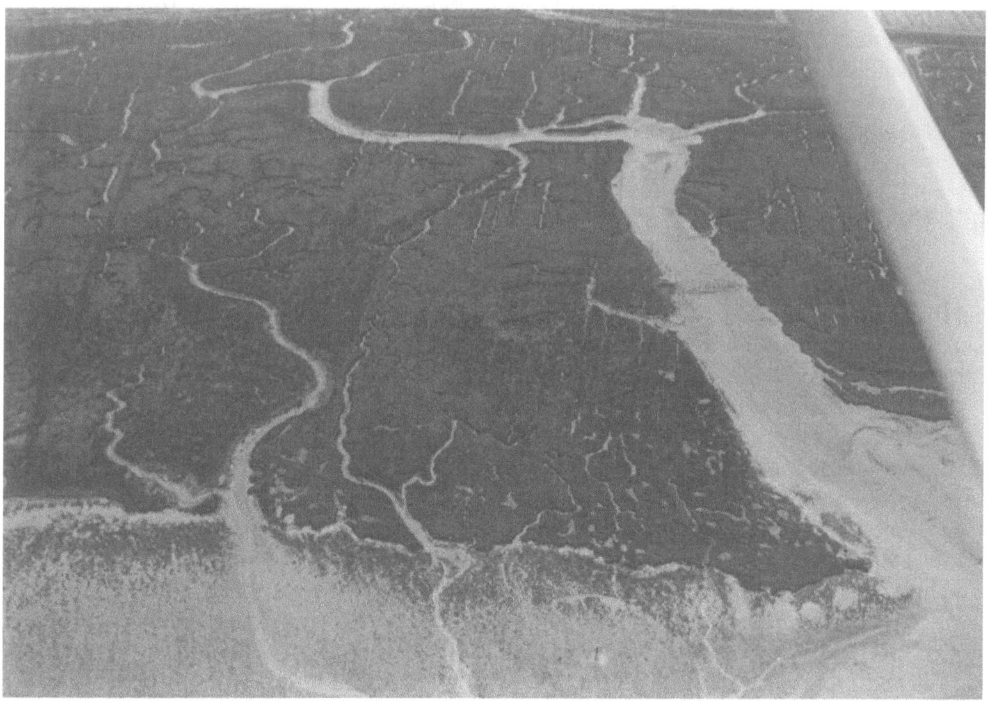

Figure 13. Not everyone knows that the fresh water gradient in estuaries results in different types of vegetation on mudflats or salt marshes.

The importance of information to the public is already emphasised in this chapter. Spaulding and Kipp (1987, pp.291-295) distinguish two basic programme types. On the one extreme a public information programme characterized by the translation and dissemination of information through media channels; on the other extreme a public participation programme characterized by various participating activities. A strong public information programme is required when a public constituency of environmentalists and users is already committed to the protection of an estuary. When the public is much less aware of pollution problems in an estuary, emphasis should be placed on education and participation rather than on dissemination of information.

Once information has been made available, the next stage of the communication process requires recipients of information to understand it and respond effectively, requiring a basic appreciation of how the public input process works and what is a useful response. A major problem here is the expectation of many respondents that their viewpoint should be taken up unchanged, and their subsequent attitude that the process is a waste of time if their view is not adopted. Public input processes should expose different user groups to each other to hear opposing points of view, rather than making it appear that each group is arguing with governmental managers. The tradeoffs between users should be made clearer to all groups.

It must be recognised that communication takes a great deal of time and resources. It is an expert field which should not be tackled by environmental managers at the end of a project, but requires the advice and services of trained personnel to determine the most effective media for a particular project and its needs. Information and public involvement should not stop with the approval of a project, but should continue in the long term, using the resources of interest groups to help with the ongoing management and maintaining their interest in protection of valuable areas.

An example of the way in which often costly external communication was dealt with is found in the project Western Scheldt Studies (Scheele and Van Westen, 1990, p.499). In this project a minimum level of information distribution was provided for by a newsletter. This newsletter provided best control of the contents of published results. As contributions were already intendedly edited with the public at large in mind, newspapers were opt to literally cite the newsletter reports.

3.7. INTERNATIONAL COOPERATION

Pollution is becoming increasingly an international issue, both in terms of its effects and the action being taken to deal with or to forestall it. While anyone in the field of estuarine management will have to work within the framework of local legislation, this is more and more being shaped by international agreements (Wilson, 1988, p.168).

Problems in estuaries are in many cases of an international character. As examples of international issues which are of importance to the developments in estuarine areas can be named: flood protection, river sediments, waste dumping, tourism, sea level rising, transport, fishing.

There have been several international conventions which have improved directly or indirectly the management of estuarine areas. These include: The Ramsar Convention on Wetlands (1971); The Oslo Convention (1972) on dumping from ships; The Paris Convention (1974) on discharges from land-based sources; The Helsinki Convention (1974) on protection of the Baltic Sea; The London Dumping Convention (1975) establishing a system to control and record dumping; The Barcelona Convention (1976) on the protection of the Mediterranean; The Convention on the Law of the Sea (1982); The Convention for the Protection of the Natural Resources and Environment of the South Pacific Region (SPREP Convention).

The effectiveness of these agreements for the improvement of the situation in estuarine areas is open to doubt. An example of the situation in Great-Britain (Rothwell and Housden, 1990, p.22) shows that the impact of international conventions on the designation of protected areas is rather limited. Firstly, we find that national authorities who are responsible for identifying sites suitable for the designation as international protected areas have only brought about a slow progress towards listing these areas. Secondly, it appears that designation on the basis of local planning rules does not always reflect the full requirements of the international convention. The provisions of special acts and planning consents on national or regional level still lead to damaging impacts. Thus, designations themselves once applied have also proved to be no guarantee for protection.

While at this stage international conventions lack an internationally enforceable implementation mechanism, and therefore have to rely on member states to put their implementation into effect, they are nevertheless useful in promoting international cooperation in various ways (OECD, 1991, p.79): They set specific reduction objectives for various pollutants which act as incentives for governments to take actions;

- Some of them were designed as comprehensive agreements for the management of specific coastal resources (e.g Helsinki and Barcelona Conventions);
- The conventions have enabled a valuable network to be established;
- The current activities can be regarded as component for integrated management on which incremental changes can be made, leading towards a more comprehensive coastal management;
- They exert moral pressure on contracting governments through political commitment to the agreement.

Conventions that focus explicitly on international coastal zone management are rare. This could be because adequate solutions have not yet been found to many of the problems, or because of the inevitable time lag between implementing policy change and the arrival of observable improvements in coastal zone areas (OECD, 1991, p.46).

Experience based on regional, inter-governmental programmes for coastal management proved the regional approach to be viable and effective. Regional action plans and coordinating procedures can stimulate capacity building through cooperation among countries sharing similar problems. Plans and procedure also help to built networks to identify regional priorities and actions be implemented and can contribute to technology transfer and training (OECD, 1991, pp.79-80).

3.8. CONCLUSIONS

In literature on estuarine management a loud call for action to improve the state of estuarine areas emerges. A great number of studies shows that the ecological situation of these coastal regions is worsening due to the many activities that have put the system under pressure. These circumstances have generated world-wide attention for ways to improve the adjustment of conflicting functions of estuarine areas. However, a growing attention for the process of estuarine deterioration coincides with a rather limited experience of estuarine management.

A major obstacle in preventing further decline of the estuarine ecosystems is the lack of coordination among the organizations involved in estuarine management. This dictates the necessity to come to a clear policy formulation in which agreement is reached among the participants in the planning process.

Planning an estuary first of all requires clearly stated goals, based on a pragmatic approach which takes into account the functions of the estuary with immediate social

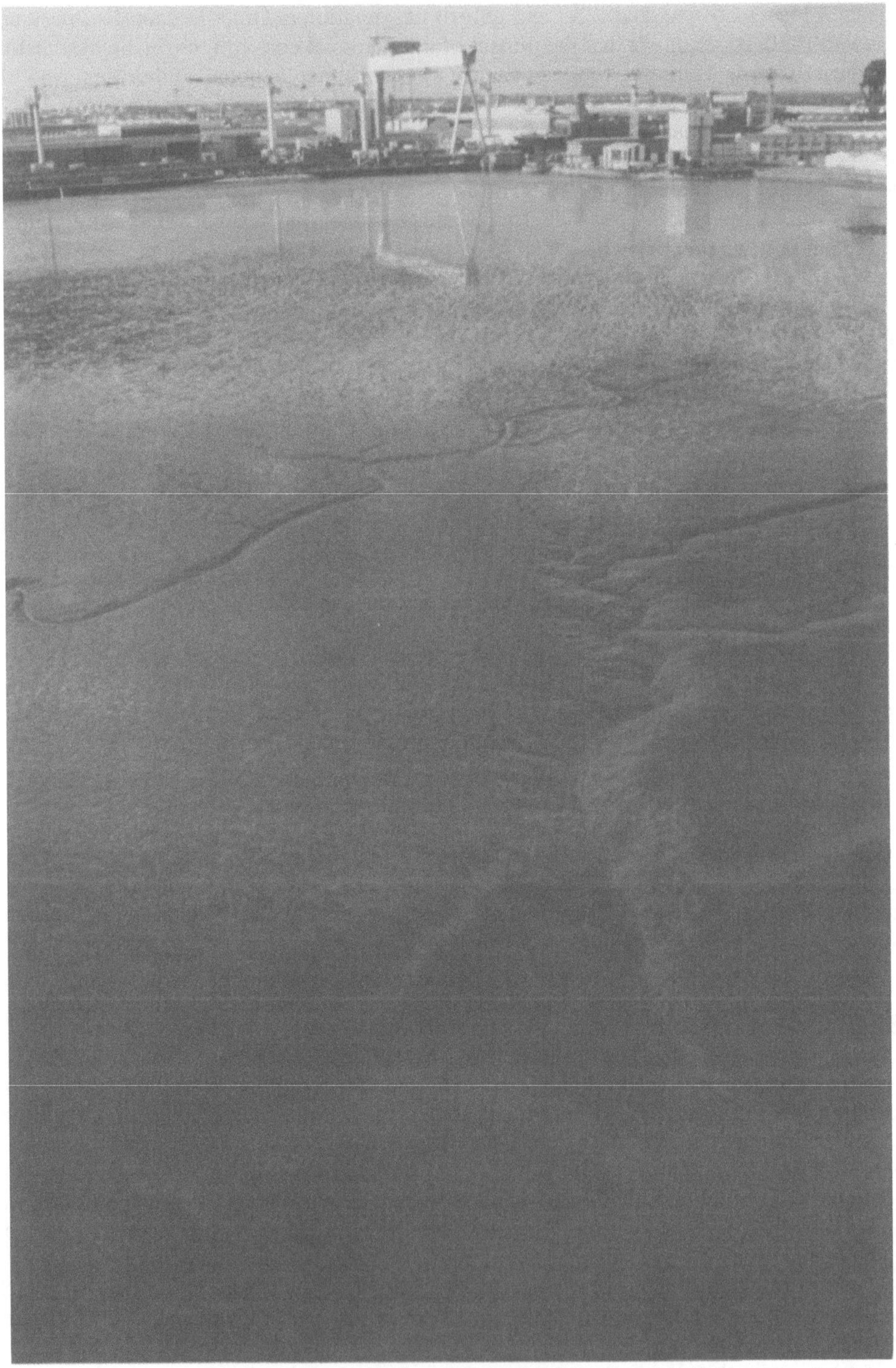

Figure 14. Industrial activities along the Loire border.

relevance. These policy aims can be more precisely defined through the adoption of environmental standards. The decision on the standards that are required is essentially a political one, the role of scientists is to advise on this matter.

To work out the objectives and standards into realistic planning options it is necessary to create an institutional mechanism in which the organizations involved in the management process are brought together. This does not necessarily entail the set up of a new body, but at least requires active coordination work of authorities on regional, national and international level.

To implement a policy, the accurate targeting of the appropriate instruments is important for success. This requires again a sound horizontal and vertical coordination of policies.

Throughout the policy process the supply of information, covering technical and natural sciences as well as economic and social fields needs full attention. Though exact information is the starting-point of correct action, in practice the impossibility of obtaining full certainty on changes in the estuarine system should not be a restraint to undertake action.

During this study it became clear that socio-economic aspects of estuarine restoration have only gained relatively little attention in the past. Publications on estuarine management show that the greater part of research work was concentrated on the biological, chemical and physical aspects of the estuarine system.

First of all this becomes apparent when we examine the amount of literature in which estuarine problems are observed from an economic and social point-of-view. A very limited number of publications that highlights estuaries from these angles. Studies that do so, originate from a relative small number of research institutes or government agencies and relate to a selected number of estuaries. This reflects the situation in certain countries (e.g. USA, Great-Britain, the Netherlands) having rather substantial experience with estuarine management, while elsewhere scientists and politicians only recently became aware of the problems.

When the existing studies on estuarine management are examined, it appears that social and economic aspects constitute most of the time only a small part of it. In most cases the attention of scientists focusses on the ecological processes while management aspects are dealt with superficially. This of course is a result of the majority of scientists with a biological, chemical or physical background working in estuarine research rather than scientists with expertise in social and economic fields.

Another important indicator for the lack of knowledge on estuarine management lies in the character of most studies. We find that a majority of the existing works only give a description of the present situation in a certain estuary. Though these studies give good insight in local situations, they do not contribute directly to an overall view on estuarine management. Of course the great diversity of estuarine areas in the world and the difference in institutional arrangements make it difficult to arrive at such a synthetic view, but greater attention for the comparison of different systems would certainly be a contribution to a better understanding of ways to improve estuarine management.

Moreover, it has become apparent that only very few studies examine the results of implemented management strategies. Due to a lack of experience studies often cannot give sufficient insight in the consequences of certain policy measures nor do they give possible ways to improve the present management strategies.

In view of the fact that estuarine management (as coordination of all the activities in estuarine areas) has gained little attention in scientific literature, special attention to this subject seems justified.

This could be done by making a more extensive comparison of overall management strategies in different estuaries. A good start here has been offered by the work of CCMS and OECD. A step further would be the setting up of a model allowing authorities involved to compare local conditions with some basic guidelines for estuarine management. Of importance is that such planning guidelines go beyond the existing general description of abstract planning rules. Useful guidelines should comprise very specific information about the ways to set up an organization and the character of measures to be taken. Above all better insight should be provided concerning the consequences of specific actions.

4

GUIDELINES

4.1. CONSIDERING SOME PRAGMATICS

It would be convenient if one could, once and for all, formulate a bunch of recipes for the proper ecological management of estuaries in any part of the world. However, every society has developed its own ways of managing problems. Especially in its legislation, some societies will often implicitly express their views how things should be handled, or, very explicitly, how they should not be handled. In other societies problems will be faced only at the time they emerge. Procedures may depend heavily on legal statements in one society, while in the other they are elaborated afterwards. In some societies comprehensive plans will be worked out, dealing with every issue at stake. In others, a simple guideline is used to work out effective management. So, a recipe, which would suit all is not possible. Yet, one could try to stipulate general ways of approaching problems, which in any society will be recognized in some way or the other by most people.

In this context we will refer to planning procedures which have been developed in the Netherlands concerning water systems. This is far from stating that these are the only available or the best; what one can say about them is that they have proved to survive practical application. Actually, the sequence of planning activities has not been as elegant as suggested. The stages mentioned are likely to parallel each other partly during their development.

The procedure for drawing up the management plans consisted of the following 13 stages, with a major in-between evaluation described in stage 8:

4.1.1. Planning Stages

1. Setting up the institutional framework
2. Setting the boundaries of the area covered by the plan
3. Assessment of the existing situation and ways in which the area is currently being used
4. Assessment of how the area would be expected to develop without change of policy
5. Assessment of the potentials of the ecosystem and the requirements of the various users
6. Evaluation of ways in which the area is used and how these uses are changing, in relation to other uses and other factors, to obtain information on actual or potential areas of conflict and to identify areas in which more information is needed
7. Formulation of aims and ways in which the uses of the area should be developed
8. Making a selection from these options and deciding on the form of management required to achieve them

9. Identifying the administrative and legal instruments required by this form of management, and developing them if necessary
10. Charting the consequences of the policy for the policies of the various regulatory agencies active in the area
11. Drawing up an action plan containing the objectives to be achieved by the various agencies during the planning period
12. Formulating a procedure for modifying the plan
13. Formulating an interim policy to be implemented during the period that the estuary is undergoing substantial changes

Setting Up the Institutional Framework. Approaches towards perceived problems may be triggered by a variety of actors. It is not always politics that initiates action. Many times private members of society or scientific contributions will start the process. However, it is crucial in modern society that problems get a place on the political agenda. This can only be achieved by proper awareness of society at large that something is wrong. Usually, one needs the support of an alert press system, to trigger action. But, above all, the political will to intervene determines potential success. Therefore, it is not always necessary to back up planning action with definite analyses. Yet, scientific findings are most welcome in convincing politics and society that appropriate measures have to be taken.

It should also be noted that the weight of the issue on the agenda has shown to change rather violently through the years. Monitoring actions by all parties involved are needed to keep people informed on progress made, or worse, deterioration going on. In order to achieve continuity in planning one needs an institutional framework that will guarantee the progress of planning activities.

Interests in society do not always support effective action, on the contrary, one may face considerable difficulties. Plans may be very hard to lift off the ground. A plan on the cupboard will not do any job, other than possibly easing the mind of those who do not want to make them work. If implementation is hard to achieve, a useful approach is making a time scale and subsequently selecting the objectives in the plan that can be met with the least effort involved, leaving the hard nuts to be cracked in the future. In most instances, the implementation of a plan demands for negotiations with a number of parties involved. Thus, at the least, someone should be made full time responsible, performing the job effectively, if possible supported by a small staff.

One could discuss at length the necessity of a larger department (which may be trustworthy, productive and stable over time, but which also may tend to insertion in the long run), or, contrariwise, a project management agency (depending on a sufficient annual budget for contract research, but vulnerable to budget changes). Either way can bring forward the necessary support, be it that there seems to be a general tendency to move towards the second option, because of alertness to problem changes and sometimes better value-for-money control.

Although most of the following may be evident, one should remind some essential considerations on the tasks to be performed. First of all one needs a clear strategy, preferably expressed in one sentence, that is supported by the major parties involved. This strategy should be broken down into explicit goals that are to be achieved. All of these goals should, whenever possible, be quantified, so as to be able to monitor the progress.

Then, 'networking' must be performed to its utmost, in order to get all the strategic information needed and to inform all parties involved. Therefore, all major actors in or affecting the estuary, be it private or public, be it active citizens or researchers, should be known, preferably on a personal basis. At least one condition is a prerequisite for 'networking': in order to achieve a recognised position outstanding knowledge of all the interests involved is an essential ingredient.

Concluding this issue: in the Netherlands, the provincial authorities play a central role in the area of physical planning. The province chairs the administrative consultation forum in which representatives from central government, local authorities and water and port authorities also participate. Within this framework, most executive activities are carried out by a core group, consisting of representatives from the provincial and central tiers of government.

Setting the Boundaries of the Area Covered by the Plan. Estuaries cannot be disconnected from their environment, and cannot be studied and managed in complete isolation. On the contrary, by their very nature of river outlet, they are subject to events that may occur hundreds of kilometres elsewhere. Also, dependent on its general geographical location, in the estuary itself, many activities may develop because of its sheltered water edge situation in between economies abroad on the one hand, and its hinterland on the other. In the early days, settlement would start near the water side, because of navigable accessibility. Nowadays, the abundant availability of water for cooling purposes or disposal of waste is, as such, a major attraction factor, whatever one may think of these uses.

For the reasons given, the temptation for development along a river and the borders of its well-situated estuary areas is overwhelming. The attractiveness of these estuaries for industries as well as tourism, almost seems to guarantee economic inflow; that is, as soon as access facilities, such as jetties, harbours and/or recreational accommodations are taken care of. In the beginning this may seem innocent and even necessary to trigger development and bring prosperity.

However, it is the danger of subsequent creeping developments, still based on disregard of their externalities, that brings a major threat to rivers and estuaries. It is not unusual and quite understandable that authorities in these locations very much welcome development - creating new employment, without putting too many constraints on entrepreneurs. Actually, each of them may even consider seriously the environmental effects and may conclude that these are not substantial. In economic terms, they consider these effects in terms of average (external) costs, not in marginal (external) costs, which they should do as soon as external effects in the basin as a whole grow substantial.

A two level approach in research as well as in policy is necessary to fully cope with the issues at stake. A two-stage research programme, then, may be a sound option to fully analyze the ecological situation of the estuary, i.e. a general apprehension stage concerning its socio-economic environment and a problem-oriented second stage dealing with the negative consequences of it. Because of the major impact of the economic status of the river basin on the estuary, a first research approach (and it may sound rather odd in ecological terms) should be aimed at a comprehensive study of the socio-economic and political activity system of the river basin. It implies that, for instance, economic geographers undertake from the very start studies in a multi-disciplinary research project of the estuary. Only, when the amount of people depending on the river for water intake and disposal of waste is known, when the type and prospects of industries and the amount of them directly using the water system are clear, when the location and capacities of waste treatment facilities in use are listed, etc., one can successfully pursue further research projects and the formulation of proper measures to be taken.

Assessment of the Existing Situation and Ways in Which the Area Is Currently Being Used. The first research approach will usually be referred to as making a regional analysis, using existing data, next to interviewing experts, thus producing a status report of the estuary. Regional analysis of estuaries is based on familiar techniques, such as time series in tables and diagrams, population pyramids showing age distribution, etc. The application of the techniques is rather straightforward, but may offer clear insights in demographic, economic

Figure 15. Typical tidal channel of the parallel coastal lagoons of Marismas Nacionales. The black mangrove (Avicennia nitida) dominates.

and other developments. Especially, when indices are presented, referring to strengths and weaknesses in the basin. Most of the time the estuary figures will in some way be compared to a larger entity in order to come to rational conclusions. This entity may be a larger region the river and estuary basin belongs to, or it may be the country as a whole. The choice of the reference region largely depends on the problem. If one wants to highlight the river and estuary basin in general, the reference region might well be the country as a whole, while, for instance, in the case development policies imply a certain amount of ecological choice between estuaries, it may make more sense to interrelate two or more estuaries, as has been done for fisheries, between comparable fisheries in the polluted Western Scheldt and those in the rather clean Eastern Scheldt.

Data to be processed in the regional analysis concern, for instance, population development, population by age, migration flows; labour force, activity rate, unemployment, gross value added, employees per economic sector, such as agriculture, industry, and services; transport figures (road, sea and air); living standards (indexes); housing (type, size, age); education level; health provisions, etc. In the case of the river and estuary basin themselves ready data may be lacking and have to be collected especially.

More and more, physical data are added to the socio-economic items, according to functions that are geographically present. Even if they are present now in some way, most of the time they have to be updated at least. Data of this kind may relate to human activities, such as dwellings, particular forms of industry, intensive or extensive agriculture, services, traffic, or nature functions, often distinguished in types of vegetation, wildlife, etc. So, all functions are mapped. On the basis of such an investigation policies are conceived of explicit zoning and put into the informal and formal decision process. The use of Geographic Information Systems is now rapidly growing, especially in the field of nature and environmental protection.

Depending on the developments traced and the problems that have emerged from the foregoing research activities, research-in-depth may follow.

Assessment of How the Area Would Be Expected to Develop Without Change of Policy. From the data processed in the first research stage one can distill the likely developments if no intervention is foreseen. By simply graphically extending developments shown in diagrams to some point in the future, one can arrive at a rough idea of future developments. Because of its ease of use and clarity of presentation, this method is widely favoured. One may, however, overlook the contribution of individual components, which make up the entire process.

More sophisticated techniques which try to incorporate to a greater or lesser extent the individual components (although not necessarily more trustworthy) are extensively presented in planning text books. They range from rather simple manual calculation algorithms, to computer-based mathematical models.

Assessment of the Potentials of the Ecosystem and the Requirements of the Various Users. A second research stage will usually be considered necessary, because of problems that emerged from the first research stage. A second stage may also be triggered if particular developments are planned, interventions are needed and effects need to be evaluated. For instance, during the last decades much research has been undertaken in determining the amount of pollution in water systems. As a result, abundant research outcomes on physical, chemical and biological contamination have become available now, be it that they may be rather fragmented, of a descriptive nature, sometimes difficult to assess and not always put in the language preferred. Carefully 'translated' to the own situation, this research, and the plans that are based on it, may provide a tentative basis for action, even if the own research capacity is very restricted, i.e. action does not have to wait for ample research to be carried out. On the other hand, it goes without saying that, whenever possible, backing up research projects should evaluate the approach, and support or reroute the course of action chosen.

Evaluation of Ways in Which the Area Is Used and How These Uses Are Changing, in Relation to Other Uses and Other Factors, to Obtain Information on Actual or Potential Areas of Conflict and to Identify Areas in Which More Information Is Needed. From the foregoing, functions of the estuary will emerge, and they will range from economic activities to nature. A simple but effective way of dealing with these issues is making a so-called conflict matrix. If one has identified the functions (such as navigation, fishing, nature, emission, cooling water) one may put these on the axes of a matrix, analyzing whether or not these functions are in conflict with each other.

Formulation of Aims and Ways in Which the Uses of the Area Should Be Developed. An aspect that deserves attention is the contact between the core group and the administrative consultation forum (see stage 1). In the cases of the three policy plans drawn up for the southwest Netherlands, stages 1 to 7 were completed in a relatively short time, after which a policy was selected and presented to the administrative consultation forum. This considerably reduced the need for further research activities.

The formulation of aims and ways in which the estuary uses should be developed depends very much on the general options of and approaches toward control. A 5-point scale of management may be useful to highlight possible differences in control (Scheele, 1991):

1. full control;
2. rigid control;
3. standard control;

4. marginal control;
5. lack of control.

Firstly, the general characteristics of the levels are discussed; secondly, the implications are highlighted.

The Full Control Situation. Full outside control towards estuaries is possible only if no one lives there to counterbalance external influences. The decision is taken on a national level, usually not without some public discussion.

The policy consequences can be summarized as: leave the estuary to nature; do not try to interfere, even if it changes of natural situation; even stop any preservation measure such as strengthening the coastline. As such, paradoxically, full control may imply no intervention at all.

The Rigid Control Situation. One may speak of a rigid control situation if the estuary faces regulations, which are not common to the national situation. Measures are actually taken by the local authorities, but they are likely to make use of supralocal provisions.

The population will be stable or declining. As a consequence the infrastructure is kept to a minimum. It only suits the restricted use by local people and some military; a minimum of public transport is provided, if any.

The Standard Control Situation. In most of the estuaries one will meet a more close-to-normal situation. May be more than elsewhere, the Netherlands policy controls land use. For instance, coastline management is an integral part of planning in general. As such the 'standard' control situation might be considered rather rigid compared to other countries.

The balanced estuary state has mainly characteristics which see economy and environment in a reasonable balance. One might define standard control situation as the economy being more or less in equilibrium with the environment. At least it is the situation strived for.

The Marginal Control Situation. Less control may be found in nations with a rather liberal political system. Market forces may then be the almost sole control mechanism, but usually some outside control is more and more felt. For instance, fisheries may be subject to international agreements.

In a marginal control situation the economy may outrule the environment. Actually, the economy is largely controlled from the outside and likely to be of a rather monolithic nature.

The Out of Control Situation. Sometimes, control may actually be absent. Poor countries, having vast territories, may find it very hard to control remote areas, if not those near-by. Poverty tends to be connected to a lack of industrialization, so the ironic statement might be that pollution is a missing asset too. Great dangers may occur, however, if polluting industries are exported from richer countries.

The main characteristics of the developments are the economic prevalence; the environment is largely ignored, although now and then some lip service is paid to satisfy worried people. As a consequence the estuary is very much open to external changes and interventions, be it economic recession or casual interference from outside controlling bodies. In all of these situations different planning aims and methods will emerge from the planning process.

Making a Selection from These Options and Deciding on the Form of Management Required to Achieve Them. Some recommendations concerning the selection of options:

Concerning the Nature of Policy Analysis. It should be evident that the analyses made are supporting decision making, and not more than that. The eventual decision is a matter of those legally responsible. Research cannot and should not pursue this role. Argumentation in this direction should be avoided.

Concerning the Contents of Policy Analysis. In short, policy analysis deals with:

- making an inventory of the actual situation and the use of land and water
- reassessment of present policies and developments
- reassessment of potential developments and interests
- definition of conflicts
- generation of alternative options and impact assessment

Concerning the Stages of Policy Analysis. Planning practice reveals that it is not always necessary to pursue integrally and in detail all stages of the analysis and evaluation process. An extensive and intensive methodological approach may turn against itself when trivial problems get overattention.

A problem will not be solved either, only by going step-by-step through the entire process. The procedure suggested is a model of reality, and any model of reality will be a

Figure 16. To further the interests of trade and shipping, many estuaries are being dredged. This human intervention often disturbs natural processes of sedimentation and erosion and harms marine life.

simplification. Therefore, occurrence of contingencies in the planning process are more likely than not.

Concerning the Selection Process. The selection process may reveal shortcomings in the alternatives. In some cases an alternative will have to be changed in order to overcome these problems. Also the discussion of constraints already agreed upon, may lead to the conception of new alternatives. This way feed back loops to earlier stages of the procedure are run through.

Concerning the Issues to be Considered. It is important not to consider too many issues. On the one hand, it may take too much time to go into detailed calculations and studies, on the other the problem of comparison of the alternatives has proven to be feasible. Clustering into groups of effects to be considered may reduce these drawbacks. In this case it makes sense to cluster effects according to the function they are related to, and put weights on them within this function frame. As a maximum number of aspects (functions) to be considered, the amount of 7 has proved to be feasible.

Concerning the Need of a Policy Plan. An estuary having several uses but not being managed will usually face serious problems, which could be coped with in a policy plan. Usually, experts will indicate priorities and measures to be taken. Doing something is always better than doing nothing.

A policy plan may be conceived by the Authorities or by private consultants. In any case, one should be aware that it is more worthwhile to have problems identified by citizens living nearby than by public officers some 100 km's away.

Concerning Socio-Economic Data Availability. The data availability may vary greatly from country to country. Especially, on regions, data may be lacking completely. In that case, one may have to conceive assessments, using higher level data.

Identifying the Administrative and Legal Instruments Required by This Form of Management, and Developing Them If Necessary. Policy development also includes related research; in order to gain comprehensive insight one needs knowledge of laws and administrative regulations that are specifically made to suit estuary/coastal problems, or knowledge of general laws that are applicable to estuaries.

On the other hand, proper policies will normally have to be supported by a broad bunch of knowledge on the subject in general. Document analysis, next to interviews of all parties involved will provide a large part of the information needed.

Charting the Consequences of the Policy for the Policies of the Various Regulatory Agencies Active in the Area. One of the most powerful instruments in the Netherlands in favour of the environment is legislation and jurisdiction on land use. Local authorities make land use plans, ultimately covering their entire territory, for a period of ten years. The plans have to be approved by regional authorities (provinces). Deviations from the plans have to be approved by the province, and they must result in a renewed, research-based plan. The making of land use plans is a process that is surrounded by extensive public debate, assuring that all parties involved are well aware of its consequences.

The main result of *zoning* safeguards to a large extent the environmental values in the Netherlands. Other areas of legislation and administration may regard garbage disposal, water quality and protection of nature.

Whenever significant policies, be it at the national, regional or local level, are formulated the regular procedure is to assign a rather comprehensive study of the problem.

In terms of research approaches, then, it is interesting to see what seems to be general practice in the field.

Methodologically speaking, the policy approach resembles more closely the linear programming approach than a simulation model approach, i.e. policies are largely constraining private behaviour. The underlying research model, therefore, might better be referred to as 'multi-disciplinary' rather than 'inter-disciplinary'. Apart from rather rare public-private projects, and subsidizing particular initiatives, public policies in general tend to be of a constraining mode, restricting adverse behaviour by zoning, legal rules, and administrative measures. This entails more and more measures based on expected behaviour rather than proven behaviour.

Therefore, for the time being, in practice it may be found more important, for instance, to use Geographic Information Systems for mapping environmental vulnerabilities, than to try and sophisticatedly model economy-environmental relationships. Although the latter might ultimately, and hopefully in the short term, provide better principles for development and protection, the present state of the art does not seem to convince politicians and pressure groups to take chances.

Drawing Up an Action Plan Containing the Objectives to be Achieved by the Various Agencies during the Planning Period. Conceiving the plan resulting from the foregoing efforts, is rather straightforward. It implies translation of research and policy analysis results into a practical packages of guidelines.

Therefore, the action plan depends very much on the insights gained by research efforts relating them to policies opted for. So, it is necessary to have some mechanism that, from the start on, synchronizes research and policy developments.

In the Netherlands the role played by research institutes in the planning process was the following. In practice, the core group set up research teams in which the planners also participated. This combination of researchers and planners contributed to the fact that research was more policy-oriented and that planners could more easily include the research results in the planning process.

Formulating a Procedure for Modifying the Plan. No plan will last forever. Not in the first place, because the plan itself may bring about fundamental changes itself, creating a necessity of change in policies. The plan may also have failed to produce proper results in one or more areas. But not the least, society may have undergone great changes, which trigger the necessity of reviewing the plan. Therefore, and this is usually laid down in formal rules such as legislation, procedures concerning the revision of the plan should be stipulated.

Formulating an Interim Policy to be Implemented during the Period That the Estuary is Undergoing Substantial Changes. It is necessary to emphasise the importance of drawing up an interim policy if the period in question is approximately 6 months or more. Every effort should be made to ensure that nothing occurs during this period, which cannot be responded to adequately.

The old Churchill rule that a plan is only good if it can be modified, should be an ingredient of the planning process. It is daily practice that plans are being changed, without affecting the quality of implementation. On the contrary, pursuing guidelines without the option to incorporate new information or conditions is detrimental to any planning process.

So, there should be ample consideration of the nature of actions to be taken in case of an emerging necessity of plan change at some point in time. Especially, responsibilities should be clear on these matters, while one should also be aware of information to be distributed concerning contingencies.

Figure 17. The storm surge barrier in the Eastern Scheldt closes only when high water levels are expected. Normally, the tides are free to enter this tidal basin. During several stages of the completion period of the barrier (1985-1987), the tides were reduced more as they are at present. An Interim Policy was formulated to reduce negative effects.

A major reason for having a provisional plan from the very start on emerges from undesirable side effects planning procedures itself may create. Well-known are price effects due to speculation and detrimental investment decisions, but even safety may be at risk, for instance, when dike-segments are known to be weak and should (even provisionally) be reinforced. Also, public awareness usually enters rather late in the planning process - when plans are becoming detailed, very much retarding the subsequent steps. A provisional plan clarifies in an early stage the major issues at stake; whoever is involved - their alert participation will be adequate.

Evidently, provisional plans should not be based on one-sided interests. Whenever feasible, scientists, experts and interest groups should be invited to state their views on no more than 3 A4's; experience shows it only takes a week or two to collect basic (not already 'politicised') insights. Yet, it will prevent major faults and enter durability into the provisional plan. This way suboptimal plans and huge redressing costs, now common to many decisions in the past, are avoided. It should leave open freedom of choice during the actual planning process.

4.2. AGENCY ACTIVITIES

To conclude this chapter on guidelines one may summarize characteristic activities, performed by agencies dealing with estuary issues. Not considering the actual organisation of an agency as this will depend very much on the culture of any country, one can consider tasks which, in any organisation will need to a certain extent attention. What activities can, besides regular organisational duties, more or less be defined as standard agency activities? A small, not necessarily exhaustive, listing may indicate tasks to be performed:

Concerning the reconnaissance of the estuary problem one may consider to:

- programme research
- assign external projects
- implement structured interviews of political actors
- scrutinize newspapers, bulletins, etc. trying to isolate major issues
- implement structured interviews of scientists
- make a list of expected strengths and weaknesses

Concerning the formulation of (general) goals regarding the estuary to:

- initiate management
- file information on all people and organizations involved in the estuary
- scrutinize past and current actions
- list all current procedures involved

Concerning the derivation of (quantified) objectives to:

- design a functions/conflicts matrix
- add to existing procedures
- define actions that can not be met by existing procedures
- list (provisional-realistic) maxima of pollution values allowed to be emitted
- list (provisional-realistic) minima of nature values to be arrived at
- define economic/social/cultural constraints

Concerning the design of alternative courses of action to:

- make function maps of the estuary basin (on a periodic basis; preferably in GIS) (i.e. a map of streams, infrastructure, industries, energy production, agriculture, urban areas, etc.)
- locate emission points of pollution
- trace areas downstream of outlets having low water velocity (such as coastal bays)
- locate vulnerable areas such as mud flats
- develop courses of action that aim at achieving the stated maxima of the objectives

Concerning the evaluation of these courses of action to:

- assign external projects
- organize a panel of experts
- provide courses for the actors involved
- produce ready information for politicians
- provide proper information for local newspapers, incidental and periodic.

5

REFERENCES

Allan, K.R.; Introduction to the Port Hacking Estuary Project. In: Cuff, W.R. and M. Tomczak Jr. (eds.); Synthesis and Modelling of Intermittent Estuaries: A case study from planning to evaluation. Springer-Verlag, Berlin etc., 1983.

Bower, B.T., C.N. Ehler and D.J. Basta; Coastal and Ocean Resources Management: A framework for analysis. IFIAS-ABC, Solna Sweden, 1982.

CCMS (Committee on the Challenges of Modern Society); Pilot study: Estuarine System Management Phase I, Final Draft Report, 1981.

CCMS; Questionnaire Characteristics and Management of Estuarine Areas; Pilot Study - Draft Report. State University Utrecht and Rijkswaterstaat, Tidal Waters Division, Middelburg, 1990.

Coccossis, H.N.; Management of Coastal Regions: the European experience. In: Nature and Resources, Vol 11, No 1, 1985, pp.20-28.

Connor, M.S.; Developing a Technical Program to Support Estuarine Management: A comparison of three northeastern estuaries. In: Lynch, M.P. and McDonald, K.L. (eds.); Estuarine and Coastal Management. Tools of Trade. Proceedings of tenth National Conference of the Coastal Society. October 12-15, 1986, New Orleans, LA. pp.503-511.

Cordosa da Silva, M. and J.M.S. Castanheiro; Estuarine Management; The Tejo-Estuary. NATO-CCMS, 1985.

Cunha, L.V.; Water Resources Management in Estuarine Systems. In: UNESCO/IOC/CNA; Estuarine Processes: An Application of the Tagus Estuary. Secrataria de Estado do Ambiente e Recursos Naturais Direccao-General da Qualidada de Ambiente. Lisbon, 1986, pp.521-539.

DeMoss, T.B.; Management Principles for Estuaries. In: Lynch, M.P. and McDonald, K.L. (eds.); Estuarine and Coastal Management. Tools of Trade. Proceedings of tenth National Conference of the Coastal Society. October 12-15, 1986, New Orleans, LA. pp.17-28.

Glasbergen, P.; Drie voorwaarden voor integraal waterbeheer (Three conditions for integrated water management). Waterschapsbelangen, no 7, 1991, pp.240-244.

Grenell, P.; Non-Regulatory Approaches to Management of Coastal Resources and Development in San Francisco Bay. In: Marine Pollution Bulletin, Vol 23, 1991, pp.503-507.

Grigg, N.S.; Estuarine Water Quality Management: Planning, Organizing, Monitoring and Modelling. In: Michaelis, W. (ed.); Estuarine Water Quality Management: Monitoring, Modelling and Research. Springer-Verlag, Berlin-Heidelberg, 1990, pp.3-10.

Hildebrand, L.P.; Canada's Experience with Coastal Zone Management. The Oceans Institute of Canada, Halifax, 1989.

Jensen, A.; Comparative Methods for Conflicting Resolution in Modern River and Wetland Management, Volume I: Danish Examples, 1988.

Juhasz, F.; An International Comparison of Sustainable Coastal Zone Management Policies. In: Marine Pollution Bulletin, Vol 23, 1991, pp.595-602.

Kriwoken, L.K. and M. Haward; Marine and Estuarine Protected Areas in Tasmania, Australia: The Complexities of Policy Development. In: Ocean and Shoreline Management. Vol 15, 1991, pp.143-163.

Leussen, W. and J. Dronkers; Progress in Estuarine Water Quality Modelling. In: Michaelis, W. (ed.); Estuarine Water Quality Management: Monitoring, Modelling and Research. Springer-Verlag, Berlin-Heidelberg, 1990, pp.13-24.

MacKay, D.W.; Progress in Estuarine Water Quality Management: An Overview. In: Wilson, J.G. and W. Halcrow; Estuarine Management and Quality Assessment. Plenum Press, New York-London, 1985.

pp.163-172. McLusky, D.S.; The Estuarine Ecosystem, 2nd Edition. Chapman and Hall, New York, 1989.

McKay, N.; Environmental Management of the Puget Sound. In: Marine Pollution Bulletin, Vol 23, 1991, pp. 509-512.

Monroe, M.W.; Environmental Activism in the San Francisco Bay Estuary. In: Marine Pollution Bulletin, Vol 23, 1991, pp.607-611.

OECD, Environment Directorate, Environment Committee; Report on Coastal Zone Management: Integrated policies and draft recommendation of the council on integrated zone management. Paris, 1991.

Ojo, O.; Strategy for Estuarine Water Quality Management in Developing Countries: the example of Nigeria (West-Africa). In: Michaelis, W. (ed.); Estuarine Water Quality Management: Monitoring, Modelling and Research. Springer-Verlag, Berlin-Heidelberg, 1990, pp.111-114.

Perret, W.S. and Chatry, M.F.; Coastal Louisiana: Abundant Renewable Natural Resources - In Peril. In: Bolton, H.S. and O.T. Magoon (eds.); Coastal Wetlands. American Society of Civil Engineers, New York, 1991. pp.317-331.

Rijsberman, F.R.; Eastern-Scheldt Estuary Study: A. Bridged report, 1988.

Robbins, S.; Managers and Management. In: Management Concepts and Applicants. Prentice Hall, 1988.

Rothwell, P. and S. Housden; Turning the Tide: A future for Estuaries. The Royal Society for the Protection of Birds, UK, 1990.

Saeijs, H.L.F.; Changing Estuaries: A review and new strategy for management and design in coastal engineering. Leiden, 1982.

Sanbongi, K.; New Legal Viewpoints for Development and Conservation of Enclosed Coastal Seas. In: Marine Pollution Bulletin, Vol. 23, 1991, pp.563-566.

Scharmann, L.; Management of Littoral Zones in Europe; Bridging the gap for new local and regional responsibilities? In: Littoral 1990, Comptes Rendus du 1er Symposium International de l'Association Européenne Eurocoast. Eurocoast, Marseille, 1990. pp.539-542.

Scheele, R.J.; Management Issues of the Western Scheldt. In: Smith, H.D. and A. Vallega (eds.); The Development of Integrated Sea-use Management. Routledge, London, 1991. pp. 199-208.

Scheele, R.J. and C.J. Van Westen; Multi-disciplinary Research Experiences; The Western Scheldt Case. In: Littoral 1990, Comptes Rendus du 1er Symposium International de l'Association Européenne Eurocoast. Eurocoast, Marseille, 1990. pp.497-506.

Scott, T.J.; Consultative Decision Making in Managing the Estuarine Environment; The Role of Policy Negotiation. In: Lynch, M.P. and McDonald, K.L. (eds.); Estuarine and Coastal Management. Tools of Trade. Proceedings of tenth National Conference of the Coastal Society. October 12-15, 1986, New Orleans, LA. pp.515-522.

Smith, H.D.; The regional Bases of Sea Use Management. In: Ocean and Shoreline Management, Vol 15, 1991, pp.273-282.

Spaulding, H.C. and K. Kipp; Establishing Public Participation Programs: The experience of the northeast estuary program. In: Lynch, M.P. and McDonald, K.L. (eds.); Estuarine and Coastal Management. Tools of Trade. Proceedings of tenth National Conference of the Coastal Society. October 12-15, 1986, New Orleans, LA. pp.291-295.

Sundermann, J., C.D. Kurt and Longfei, Ye; Strategy of Current and Transport Modelling in different Estuaries. In: Michaelis, W. (ed.); Estuarine Water Quality Management: Monitoring, Modelling and Research. Springer-Verlag, Berlin-Heidelberg, 1990, pp.24-34.

Ten Brink,B.J.E., Beleidsinstrumenten (Instruments for policy making), Rijkswaterstaat, Tidal Waters Division, Den Haag, 1988, 41 pp.

The Seto Inland Sea Declaration. In: Marine Pollution Bulletin, Vol 23, 1991, pp.XVII-XVIII.

Thoemke, K.; Estuarine Management at the Rookery Bay National Estuarine Research Reserve. In: Lynch, M.P. and McDonald, K.L. (eds.); Estuarine and Coastal Management. Tools of Trade. Proceedings of tenth National Conference of the Coastal Society. October 12-15, 1986, New Orleans, LA. pp.771-772.

Van Westen, C.J. and C.J.Colijn. Policy Planning in the Oosterschelde estuary. In: The Oosterschelde estuary (The Netherlands): A Case-Study of a Changing Estuary. 1994, Hydrobiologia 282/283, Kluwer Academic Publishers, Dordrecht.

Van Westen, C.J., A. Van der Wekken and B.J.W.M.Devilee. Beheersplan voor de Rijkswateren (National Management Plan for State-administered waters). Ministry for Transport and Public Works, Rijkswaterstaat. SDU, Den Haag, 1993, 263 pp.

Wilson, J.G.; The Biology of Estuarine Management. Croom Helm, London, 1988.

Winsemius, P., Gast in eigen huis. Beschouwingen over milieumanagement. (Guest in your home. Opinions about environmental management). Samsom H.D. Tjeenk Willink, Alphen aan de Rijn, 1987, 227pp.

Wood, P.C.; An Idealised System for the Management of the Estuary of the Tagus. In: UNESCO/IOC/CNA; Estuarine Processes: An Application of the Tagus Estuary. Secrataria de Estado do Ambiente e Recursos Naturais Direccao-General da Qualidada de Ambiente. Lisbon, 1986, pp.491-503.
World Commission on Environment and Development, 1987. Our common future. Oxford University Press.

INDEX

MIX
Papier aus verantwortungsvollen Quellen
Paper from responsible sources
FSC® C105338

If you have any concerns about our products,
you can contact us on
ProductSafety@springernature.com

In case Publisher is established outside the EU,
the EU authorized representative is:
Springer Nature Customer Service Center GmbH
Europaplatz 3, 69115 Heidelberg, Germany

Printed by Libri Plureos GmbH
in Hamburg, Germany